Advanced Information Processing

Springer-Verlag Berlin Heidelberg GmbH

Jiming Liu • Laeeque K. Daneshmend

Spatial Reasoning and Planning

Geometry, Mechanism, and Motion

With 82 Figures and 6 Tables

Jiming Liu
Department of Computer Science
Hong Kong Baptist University
Kowloon Tong
Hong Kong

Laeeque K. Daneshmend
Department of Mining Engineering
Goodwin Hall, Queen´s University
Kingston, Ontario K7L 3N6
Canada

Library of Congress Cataloging-in-Publication Data applied for

Die Deutsche Bibliothek - CIP-Einheitsaufnahme
Bibliographic information published by Die Deutsche Bibliothek
Die Deutsche Bibliothek lists this publication in the Deutsche Nationalbibliografie; detailed bibliographic data is available in the Internet at <http://dnb.ddb.de>.

ACM Subject Classification (1998): I.2.10, I.3.5, J.2

ISBN 978-3-642-62337-0 ISBN 978-3-642-18879-4 (eBook)
DOI 10.1007/978-3-642-18879-4

springeronline.com

Originaly published by Springer-Verlag Berlin Heidelberg New York in 2004
Softcover reprint of the hardcover 1st edition 2004

Cover Design: KünkelLopka, Heidelberg
Typesetting: Computer to film by author´s data

Printed on acid-free paper 45/3142PS 5 4 3 2 1 0

Preface

Spatial reasoning and planning has been widely applied in Geospatial Information Technologies, Computer-Aided Design, Robotics, and Automation. Some examples of application are:

1. Evaluating pre-parametric mechanism designs given qualitative geometric specifications;
2. Analyzing motion constraints for mechanism-centered robot manipulation tasks;
3. Planning general routes as well as exact paths for robots, subject to the geometric constraints of a set of obstacles in the environment.

The aim of this book is to provide a new *synthesis* of spatial reasoning and planning techniques, and to demonstrate it with various illustrative examples. Our synthesis consists of several key contributions:

1. A unified *framework* of spatial representation and reasoning is systematically constructed and presented.
2. Formalisms for the qualitative abstraction of spatial quantities, such as the Euclidean *distances* and *angles*, are constructed and spatial inference rules are accordingly formulated. Theorems concerning the minimum requirements and completeness of the spatial inferences are provided.
3. A formalism for the spatial characterization of a polygonal environment based on the notion of regions is described. Several properties of the m-closure partition of the environment are identified with respect to the uniqueness and upper and lower bounds of the regions. Based on this representation, the notions of locations and routes as well as three qualitative *optimality criteria* for measuring the length, passage clearance, and orientation cost of the route are established. Based on the definition of m-edge configurations, a new segmentation scheme is presented to represent a path within a polygonal environment.
4. The spatial inferencing techniques are applied to deriving the instantaneous configurations as well as velocity relationships of *planar linkages*. A

set of detailed algorithms for controlling the spatial propagation is formulated. The qualitative approach produces the ranges of spatial variables which can limit the possible configurations for local search (i.e., search space for simulated annealing) in generating exact configurations. This approach is suited for problems where the initial information about the metrics of the linkages is given in *qualitative* terms.

5. The techniques for planning a global qualitative route as well as an exact path given a set of connected m-closure regions, are demonstrated with *robot-like under-constrained mechanisms* in the presence of a set of static obstacles.

The book contains ten chapters:

Chapter 1 reviews previous work in related areas, highlighting the methods, implemented (experimental) systems, and arguments presented by researchers in Computer-Aided Mechanism Design, Artificial Intelligence, Robotics, and Spatial Modeling. It attempts to give a comparative overview of the earlier work as well as historical background of the present research.

Chapter 2 presents an in-depth description of the general spatial reasoning and planning problems. It begins by defining the terminologies and notations that are used throughout the rest of this book, and follows this by stating the problems in a formal manner.

Chapter 3 introduces formalisms for representing spatial relationships. These formalisms partition the continuous domains of spatial quantities in Euclidean space into mutually disjoint subdomains labeled by qualitative linguistic terms. This chapter lays down the foundation for the later discussions of spatial inferencing and planning.

Chapter 4 describes the methods of spatial inferencing and spatial planning. Both methods require an *envisionment* step for deriving possible qualitative spatial configurations and a *simulated-annealing* step for generating more exact solutions. In the spatial analysis method, the envisionment is created by propagating qualitative spatial relationships with the inference rules of *qualitative trigonometry* and *qualitative arithmetic*, whereas in the spatial planning method, the envisionment is created by using a heuristic search technique.

Chapter 5 shows how the spatial analysis method, introduced in the preceding chapter, can be applied to analyzing the motion trajectories of constrained planar mechanisms. First, it addresses the problem of instantaneous configuration envisionment. Then, it describes the generation of constrained motion trajectories that draws on the instantaneous configuration analysis.

Chapter 6 also concerns itself with one-degree-of-freedom (i.e., constrained) planar mechanisms, but focuses on the velocity aspect of the constrained motion. Here, the velocity relationship problem is geometrically treated in a way similar to the trajectory analysis.

Chapter 7 addresses the issue of how to analyze the velocity relationships of a linkage mechanism given its dimensional specifications. Two methods discussed are: (1) velocity analysis-based instantaneous axis and (2) relative motion vector analysis.

Chapter 8 deals with the application of the spatial planning method, as introduced in Chapter 5, to generate the valid configurations of many-degrees-of-freedom planar mechanisms in the presence of polygonal objects. The first three sections present an outline of the general algorithm as well as the derivation of detailed algorithms. The last three sections demonstrate the method by showing the results of software simulations, and discuss its major features and limitations. The simulation-based experiments cover both single-link and multi-link mobile object cases.

Chapter 9 focuses on the issue of spatial measurement and modeling. Specifically, it discusses two topics: (1) how to map uncertainty associated with sensing measurements to uncertainty associated with the perceptual hypotheses based on those measurements, and (2) how to collectively derive a spatial map of an unknown environment using a group of autonomous robots.

Chapter 10 concludes the book by highlighting the significant notions in spatial reasoning and planning, discussing various avenues for practical applications.

Features of the Book

This book has a number of specially-designed features that will help readers to understand the topics presented and to apply the techniques demonstrated. These features are:

1. The contents have a balanced emphasis on the pioneering work, the theoretical or computational aspects, and the practical applications of automated spatial reasoning and planning.
2. The materials are structured in a systematic, coherent manner, and hence can be readily used as basic course materials in an engineering discipline.
3. A unifying engineering-oriented approach toward the use and development of automated spatial reasoning systems is introduced with implementation details, comprehensive examples, and case studies.
4. Many graphical illustrations are inserted to explain and highlight the important notions and principles.
5. Detailed algorithms are provided for readers to develop their applications.

Intended Audiences

This book is intended as a self-contained introductory text for several categories of audience. It can be conveniently used as course material by final-year undergraduate and graduate-level students in computer science and most engineering disciplines in learning about how to develop and apply spatial reasoning models, methodologies, and tools for solving real-world computational and engineering problems.

The book is also suited to computer scientists, engineers, researchers, and practitioners in the field of intelligent machines, who are interested in finding solutions to problems arising in their development of intelligent and autonomous systems. The methodologies, algorithmic details, and case studies presented in this book can offer a convenient reference.

For instance, after reading some chapters of this book, one will be able to clearly address the following issues in his or her systems development:

1. How to map continuous spatial quantities into *qualitative representations* that enable (i) *inferences* about qualitative spatial relationships within a one-degree-of-freedom (i.e., constrained) planar mechanism, and (ii) the *synthesis* of a sequence of qualitative spatial configurations for an open-chain planar mechanism constrained by a set of polygonal objects.
2. Given geometric measurements (either qualitative or quantitative) of a one-degree-of-freedom planar mechanism with revolute or prismatic joints, how to approximately describe (i) the constrained-motion *path*, and (ii) the constrained *velocity relationships* of the mechanism.
3. Given exact geometric information about an open-chain planar mechanism (e.g., a manipulator or a mobile robot) with or without kinematic joints, a set of polygonal objects in the environment, and the relative spatial relationships between the mechanism and its environment, how to derive the *path* of the mechanism which geometrically satisfies the constraints imposed by its joints and by its environment.

Acknowledgements[1]

We want to acknowledge several sources of financial support, which have made this work possible. Centre de Recherche Informatique de Montréal (CRIM) supported the earlier phase of this work through the CRIM/McGill/Hydro-Québec telerobotics project. Additional support was provided by Project C-2 of the Canadian Institute for Robotics and Intelligent Systems (IRIS). Partial support was provided by Fonds pour la Formation de Chercheurs et l'Aide a la Recherche (FCAR), as a scholarship.

We feel privileged to be able to carry out part of this work in the Center for Intelligent Machines (CIM) at McGill University, and to be able to interact with professors and graduate students in the Center.

We would like to express our thanks to Michel Desmarais of CRIM for allocating time and resources for this work, and providing suggestions on the

[1] Omissions of credit acknowledgement in this book, if any, will be corrected in future editions.

presentation of this book. Special thanks go to Paul Freedman and Jianbing Wu for their discussions and inputs with respect to the contents and presentation in Chapter 9.

We would like to offer our gratitude to Hong Kong Baptist University and Queen's University in Canada for their support over the past ten years.

Writing and publishing a book is a grand project. The undertaking of this project becomes even more challenging in light of Jiming Liu's role as the Head of Computer Science Department at HKBU. We are very grateful to our families and friends for their strong support, encouragement, and understanding.

This project could not be carried out so smoothly and successfully without the coordination and help from Ralf Gerstner and other staff of Springer-Verlag.

Hong Kong
December 2003

Jiming Liu
Laeeque K. Daneshmend

Contents

1 Introduction

This book presents a framework for *qualitative spatial representation* and *reasoning*. The qualitative representation of spatial relationships provides a general vocabulary for describing distinctive spatial configurations as well as a set of inference rules for qualitatively reasoning about the configurations. The framework of qualitative spatial reasoning enables the generation of qualitative solutions to spatial problems where the geometric knowledge is *imprecise*. It can also provide approximate guidance to the application of quantitative methods.

The book demonstrates two applications of the qualitative spatial reasoning framework in the domains of planar mechanism analysis and motion planning. The considered mechanisms are mechanical assemblages and composed of a set of rigid bodies connected by kinematic joints, such as revolute joints (i.e., hinges) and prismatic joints (i.e., sliding joints). The mechanism motions are *geometrically* constrained by the joints in the mechanisms or by other objects in their environment. Some examples of such mechanisms are shown in Figure 1.1.

1.1 Motivation

Mechanism analysis and motion planning have direct relevance to and implications on *computer-aided mechanism design* as well as *robotics* in general. In particular, the qualitative spatial reasoning approach, as described in this book, is *complementary* to the existing mechanism-analysis and motion-planning approaches, and can find applications in the following task domains:

1. Evaluating pre-parametric mechanism designs (given qualitative geometric specifications).
2. Analyzing motion constraints for mechanism-centered robot manipulation tasks.
3. Planning general routes as well as exact paths for robots, subject to the geometric constraints of a set of obstacles in the environment.

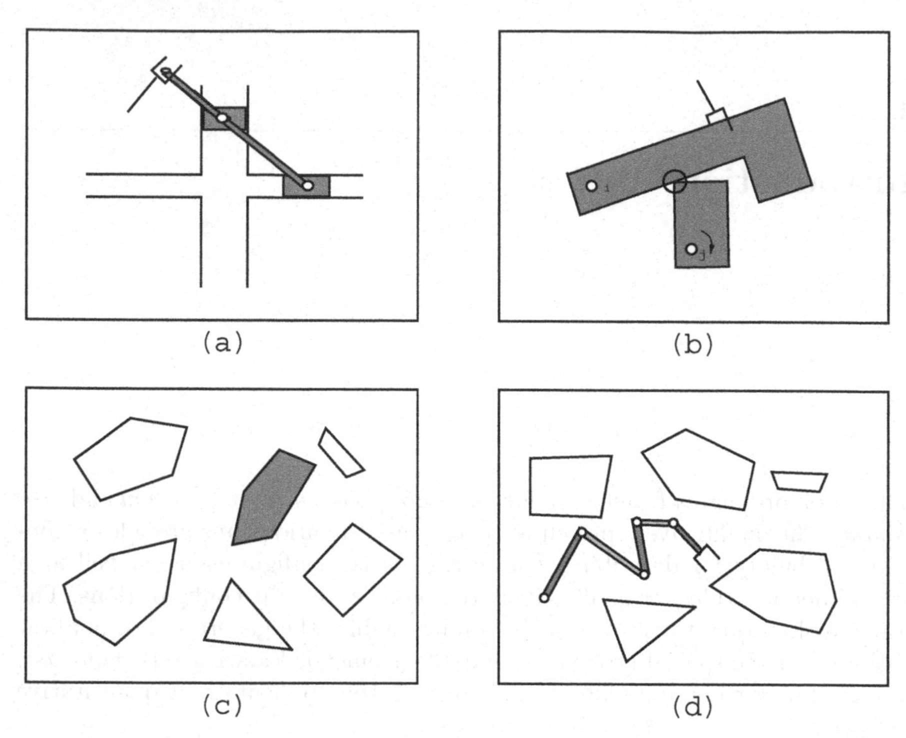

Fig. 1.1. Some examples of the mechanisms often studied in computer-aided design and in robotics. (a) A closed-chain constrained linkage mechanism. In pre-parametric mechanism design as well as in planning robot manipulation tasks, it is desirable to analyze the general *behavior and motion constraints* of a qualitatively-perceived *linkage mechanism.* (b) A mechanism with a point-contact. Some mechanisms may contain kinematic pairs (i.e., joints between individual bodies) other than hinges (i.e., revolute pairs) and sliding joints (i.e., prismatic pairs). (c) An under-constrained mechanism in a polygonal environment. In this case, the under-constrained mechanism is a single mobile rigid object. The problem of finding an optimal path has direct relevance to *mobile robot* path planning. (d) An open-chain under-constrained linkage mechanism in a polygonal environment. The problem of *manipulator* path planning can be treated as a special case of under-constrained mechanism motion planning.

In computer-aided mechanism design, analytical techniques such as standard-form and loop-closure equations have been exclusively used due to the accuracy and completeness of the techniques [170]. However, the success in using them to synthesize mechanisms is based on the assumption that the conceptual design of the mechanism provides an appropriate mechanism type for a given performance.

Post-conceptual design analysis is an intermediate step between conceptual design and dimensional synthesis. It evaluates the *qualitative* layout generated from the conceptual designs. The typical questions to ask at this stage are as follows: What general dimensions of a mechanism will be suitable for generating desired kinematic behavior? What configuration or starting position is appropriate? In order to answer these questions, simulations of the proposed mechanism based only on *qualitative spatial information* will be desirable. In such a case, a computer may be programmed to qualitatively reason about and test mechanisms of various dimensions until it obtains the desired motion characteristics.

In robotics, one of the ultimate objectives is to build robots with the capability of understanding high-level instructions. With an automatic planner and sensing capability, robots can accomplish tasks and deal with uncertainty with fewer human interventions. Such robots should be able to plan their specific motions connecting given initial and goal configurations that satisfy some spatial constraints as imposed by other objects in the environment [127]. The problem of topological and geometric path planning has been studied by many researchers. Most of the existing methods are geometrically exact [126, 129, 138, 139, 188]; some are heuristics-based [1, 93]. In this book, the motion-planning problem is viewed as a special case of path generation by a *general under-constrained mechanism.*

Mechanism motion analysis is also essential in planning robot tasks which require the manipulation of mechanical assemblages. Some examples of such tasks are closing a door or a window, releasing a cable, and turning a crank or a handle. Here, the central robot motion-planning problem is to enable a *task planner* to determine the necessary robot manipulation strategies with respect to a mechanism-oriented task specification. In planning mechanical device manipulation, models of device kinematics will facilitate the identification of the robot motions that are consistent not only with specific task requirements but also with external constraints. In fact, the problem of planning mechanism manipulation tasks can, in most cases, be reduced to that of kinematic analysis of the mechanism being manipulated [147].

With respect to the above robotics problems, the methods developed in the current work can find direct applications. In particular, the methods can be applied to characterize the motion of a constrained planar mechanism for planning robot manipulation tasks — a *compliant motion planning* problem [147] and can also be used for the geometric planning of collision-free motions for a redundant mechanism (with many degrees of freedom) within a polygonal environment — a typical *robot-path-planning* problem [128].

The methods developed in this book are *complementary* to the existing motion analysis and planning methods in that they incorporate qualitative spatial knowledge, and should have the following general thrusts:

1. They do not require complete or precise representations of geometric constraints.
2. Qualitative spatial descriptions are used to serve as a guide to the *exact motion* analysis, and the results of analyses can be explained intuitively.
3. They are computationally tractable, efficient, and easy to implement.

1.2 Issues

The issues central to this book are as follows:

1. How to map continuous spatial quantities into *qualitative representations* that enable (i) *inferences* about qualitative spatial relationships within a one-degree-of-freedom (i.e., constrained) planar mechanism, and (ii) the *synthesis* of a sequence of qualitative spatial configurations for an open-chain planar mechanism constrained by a set of polygonal objects.
2. Given geometric measurements (either qualitative or quantitative) of a one-degree-of-freedom planar mechanism with revolute or prismatic joints, how to approximately describe (i) the constrained-motion *path*, and (ii) the constrained *velocity relationships* of the mechanism.
3. Given exact geometric information about an open-chain planar mechanism (e.g., a manipulator or a mobile robot) with or without kinematic joints, a set of polygonal objects in the environment, and the relative spatial relationships between the mechanism and its environment, how to derive the *path* of the mechanism which geometrically satisfies the constraints imposed by its joints and by its environment.

The above statements will be elaborated in Chapter 3, where related terminologies are defined.

1.3 Scope of the Book

This book is concerned with the motion of *planar mechanisms* subject to geometric constraints. While investigating the trajectories and velocities of one-degree-of-freedom mechanisms and the paths of open-chain mechanisms, other aspects of the motion such as acceleration and forces acting on the mechanisms will not be considered. The latter will require fundamentally different methodologies in terms of both representation and reasoning formalisms.

In the current work, the mechanisms to be considered consist of *linkages*, which cover a significantly large class of applied mechanisms. Joskowicz and Sacks [98] have surveyed 2,500 mechanisms in a mechanical engineering encyclopedia [4], and found out that the mechanism-type distribution is the following: *linkages* (35%), fixed-axes (22%), *fixed-axes coupled by linkages* (9%), and complex mechanisms (34%).

The methods developed in this work are applicable to mechanism analysis and motion planning, and will be demonstrated through graphical simulation-based case studies. This simulation approach provides a basis for further implementation and application of the theory developed here, e.g., for computer-aided mechanism design systems and robot motion planners.

1.4 Organization of the Book

The book is organized as follows:

Chapter 2 reviews previous work in related areas, highlighting the methods, implemented (experimental) systems, and arguments presented by the researchers in Computer-Aided Mechanism Design, Artificial Intelligence, Robotics, and Spatial Modeling. It attempts to give a comparative overview of the earlier work as well as an historical background of the present research.

Chapter 3 presents an in-depth description of the problems addressed in the current work. It begins by defining the terminologies and notations that are used throughout the rest of this book, and follows this by stating the problems in a formal manner. The specific assumptions adopted in the book are also presented.

Chapter 4 introduces formalisms for representing qualitative spatial relationships. These formalisms partition the continuous domains of spatial quantities in Euclidean space into mutually disjoint subdomains labeled by qualitative linguistic terms. This chapter lays down the foundation for the later discussions of qualitative spatial inferencing and planning.

Chapter 5 describes the methods of spatial inferencing and spatial planning. Both methods require a *qualitative-envisionment* step for deriving possible qualitative spatial configurations and a *simulated-annealing* step for generating more exact solutions. In the spatial analysis method, the envisionment is created by propagating qualitative spatial relationships with the inference rules of *qualitative trigonometry* and *qualitative arithmetic*, whereas in the spatial planning method, the envisionment is created by using a heuristic search technique.

Chapter 6 shows how the spatial analysis method, introduced in the preceding chapter, can be applied to analyzing the motion trajectories of constrained planar mechanisms. First, it addresses the problem of instantaneous configuration envisionment. Then, it describes the generation of constrained motion trajectories that draws on the instantaneous configuration analysis.

Chapter 7 is also concerned with one-degree-of-freedom (i.e., constrained) planar mechanisms, but focuses on the velocity aspect of the constrained motion. Here, the velocity relationship problem is geometrically treated in a way similar to the trajectory analysis.

Chapter 8 deals with the application of the spatial planning method, as introduced in Chapter 5, to generate the valid configurations of many-degrees-of-freedom planar mechanisms in the presence of polygonal objects. The first three sections present an outline of the general algorithm as well as the derivation of detailed algorithms. The last three sections demonstrate the method by showing the results of software simulations, and discuss its major features and limitations. The simulation-based experiments cover both single-link and multi-link mobile object cases.

Chapter 9 focuses on the issue of spatial measurement and modeling. Specifically, it discusses two topics: (1) how to map uncertainty associated with sensing measurements to uncertainty associated with the perceptual hypotheses based on those measurements, and (2) how to collectively derive a spatial map of an unknown environment using a group of autonomous robots.

Chapter 10 concludes the book by highlighting the key contributions of the current work, discussing various avenues for practical application and the limitations of the proposed approach, and pointing out the directions for future investigation.

2

Overview of Spatial Reasoning and Planning Techniques

This chapter provides a review of previous work. It focuses on the themes of current interest in the following broad areas: computer-aided kinematic design of mechanisms, geometric path planning, qualitative reasoning and spatial analysis, and simulated annealing.

2.1 Computer-Aided Kinematic Design of Mechanisms

Many approaches to kinematic design of mechanisms have been developed in the past. Broadly speaking, they deal with two general problems: (1) conceptual or type design and (2) dimensional synthesis.

Research efforts on the first problem aim at identifying an array of the best possible mechanism layouts. Regarding type design as a general engineering design process, a variety of descriptive, prescriptive, and computer-based models of human design processes have been proposed [60, 177].

Yoshikawa et al. [109, 179, 180] have introduced a *general design theory* in which the notion of *metamodels* plays a crucial role in describing the evolutionary nature of design. In the metamodels, the behavior of a design object is represented in terms of qualitative processes [61, 123].

Chandrasekaran [24] argues that such design problem solving is essentially an activity of recursively selecting propose-critique-modify methods. Gero [70] and Gero and Rosenman [71] suggest that a system of prototypes be used as conceptual schema for knowledge-based design systems. Prototypes such as classes of design allow for the description of performance and structure properties. The central problem in reasoning about physical objects is reasoning about the way in which objects move and interact with each other. Kusiak [124] has described several systems that attempt to address this.

Directly concerned with mechanism type design, Duffey and Dixon [41] have proposed a model of topological and geometric reasoning about mechanical designs, and have illustrated it with the two-dimensional extrusion cross-section as a case study. Their approach is to modify topological features as well as parametric design variables with a set of operators.

A systematic approach for mechanism type design has been developed by Yang et al. [192], in which graph theory is applied. Such an integrated methodology has been demonstrated with an implemented expert system called *DOMES* (*Design Of Mechanisms by an Expert System*). This system generates sketches of mechanisms that satisfy given requirement specifications.

Other researchers have taken a heuristics-based approach to the problem of type design. Dyer and Flowers [42] and Dyer et al. [43] have introduced the idea of automating mechanical design by means of heuristics and analogy-based symbolic manipulations on representational constructs (e.g., goal or plan information, spatial relationships, forces, motion, and contact, etc.). Kota et al. [116] have extensively studied the motion characteristics of hundreds of linkages (e.g., Dwell linkages), and have developed a comprehensive *classification* system. This qualitative classification of mechanisms allows the selection of the best possible linkage design, given a set of specifications.

In addition to mechanism type design, kinematic synthesis of mechanisms has also been extensively studied. Here, the central problem is to determine the detailed dimensional design of the mechanisms so that specific performances can be achieved.

The *LINKG* program , developed by ElMaraghy and Newcombe [45] in 1977, represents one of the first *interactive* computer-aided design packages for linkage synthesis and analysis. *LINKG* can generate not only velocity and acceleration diagrams but also linkage progression and *real-time* animated mechanism motions, and hence it provides an effective and attractive way of teaching and carrying out the design process.

The commonly-used formulations for mechanism analysis are vector-loop equation and matrix manipulation, as shown by Hartenberg and Denavit [80], Hartenberg et al. [183], and Shigley and Uicker [172]. These approaches require iterations and numerical methods [47]. Other well-known approaches employ geometric techniques [90, 91] and constraint propagation [147]. Salem and Manoochehri [169] have developed a system for the kinematic and dynamic analysis of planar mechanisms based on the component approach as proposed by Suh and Radcliffe [178]. The component approach decomposes a complex planar linkage into a set of simpler components, and then sequentially solves the closed-form equations of the components. Most of the aforementioned approaches rely on precise mathematical formulation, and assume that the conceptually designed mechanisms are correct.

Kota et al. [116] argue that in order to achieve a complete computer-aided design system, a methodology that incorporates an iterative generate-and-test process should be developed. Such a process can be referred to as post-conceptual design analysis – an intermediate step between the type design and the detailed optimal design. One of the primary concerns of this process is to evaluate the *qualitative* layout generated from the conceptual design, and to save the detailed analytical synthesis from unnecessary trials and errors.

In order to support the iterative generate-and-test of mechanisms at the post-conceptual design stage, it is clearly desirable to have a means of analyzing mechanism kinematics that does not solely rely on the exact geometric information about mechanisms. This book in part addresses this issue and provides a methodology for solving the problem of constrained closed-chain mechanism analysis.

2.2 Geometric Path Planning

This section reviews the related work on geometric path planning. Although most of the studies have concentrated on the motion of robot-like mechanisms, namely single mobile objects and open-chain manipulators, the underlying principles and methodologies may also be applicable to the investigation of *general mechanism* paths under geometric constraints. These methodologies are reviewed under three topics: path search in configuration space, path finding with direct free-space characterization, and local path planning.

2.2.1 Path Search in Configuration Space

Among the earlier efforts, Udupa [182] proposed a method of motion planning by transforming a robot into a point in parameter space. Lozano-Pérez and Wesley [143] extended further the idea of parameter-space representation and formulated a *general* representation framework for computing collision-free paths. In their framework, the position of a robot relative to its environment is described in the position parameter space of the robot, known as *configuration space* or $\mathcal{C}$-space. Hence, the problem of finding a path from an initial position to a final position can be reduced to that of *searching* a subspace of the configuration space, i.e., free-space, and building a path that connects the initial and the goal (final) configurations [138]. This approach guarantees to find a path if one exists within the free-space; it is also feasible to search for short paths, rather than to simply find the first path that is safe.

In recent years, several approximate numerical algorithms for performing the $\mathcal{C}$-space search have been proposed [56, 57, 58, 140]. In addition, a number of path planners based on this approach have been implemented. Examples of such planners are *Handey* by Lozano-Pérez et al. [141] and *ACT* by Bellier et al. [8] and Mazer et al. [150]. The motion-planning techniques have also found applications in other domains, such as computer-aided pipe layout design [197].

In the $\mathcal{C}$-space approach, there are two key steps: (1) representing the robot free-space by discretizing its configuration space and (2) constructing a transfer path between two configurations by performing a search in the connectivity graph. As far as the discretization is concerned, there exist a number of methods for decomposing the free-space of the robot into a set of disjoint regions; this is also known as *cell decomposition*. These methods differ from one to another in the definition of the regions; some of them are exact decompositions, while others are approximate. Examples of the decompositions typically applied are convex polygonal decomposition [101, 102], trapezoidal decomposition [26], rectangular decomposition [19, 140, 198], 2^n-tree decomposition [56, 85], and orientation slicing [138].

All of the above methods assume that the complete information on the robot environment is known prior to planning. In order to solve the problem of planning motion in the presence of modeling, control, or sensing uncertainty, Lozano-Pérez et al. [142] have developed a formal method of automatically synthesizing fine-motion strategies, known as *preimage backchaining*. The preimage of a goal configuration is the subset of robot configurations from which, given a specific motion command, the robot is guaranteed to reach the goal. The preimage backchaining approach generates a motion plan by iteratively computing preimages backwards from the goal configuration until a preimage is found which contains the initial configuration(s). This approach provides a general framework for robot motion planning with uncertainty.

Other researchers have later contributed to the preimage approach. Erdmann [48] has proposed a *backprojection* method which distinguishes the computation of the goal reachability from that of goal recognizability. Donald [39] has formulated a world-modeling formalism, *generalized $\mathcal{C}$-space*, which takes into account the uncertainties during $\mathcal{C}$-space modeling. In addition, he has also addressed the problems of recovering from task execution errors and computing motion plans that are *likely* to succeed in spite of geometric modeling errors. Lazanas and Latombe [129] have applied the preimage approach to robot navigation. An integrated approach based on the previous methods of computing preimages have been provided by Latombe et al. [128] for a robot with two-dimensional Euclidean configuration space. This combination has resulted in a practical method for implementing fine-motion planners, as shown in their experiments. However, as pointed out in [128], their approach does not employ any heuristics in building and searching the preimage graphs. One of the possible improvements will be to incorporate qualitative knowledge such as "move with priority in the direction of the goal".

Buckley [21] has taken a slightly different approach, in which the task geometry is represented as a finite state space whose states are collections of features from the configuration space, such as *contact surfaces*.

ElMaraghy and Payandeh [46] have shown how the likely contact surfaces between convex bodies can be inferred based on the measured force vector and partial *a priori* knowledge of mating parts geometry.

To deal with modeling uncertainty in a stepwise fashion, Lumelsky [144] and Lumelsky and Stepanov [145] have introduced a Dynamic Path Planning (DPP) approach with some *nonheuristic* algorithms for robot arms of various kinematic configurations moving amidst unknown obstacles. This approach allows real-time motion planning for simple robot arms. However, it is limited only to gross position planning. Freedman and Liu [67] have provided a formalism which can be applied to model the uncertainty associated with sonar measurements.

2.2.2 Path Finding Based on Direct Free-Space Characterization

While the work on the $\mathcal{C}$-space approach focuses on search in a parameter space, other researchers have proposed methods of motion planning with an explicit characterization of the robot free-space. These methods directly decompose a robot's free workspace, without first transforming it into a $\mathcal{C}$-space representation.

Brooks [18] has proposed a method of representing two-dimensional robot free-space as natural "freeways" between obstacles, and presented an algorithm to find collision-free paths. Other representations similar to the "freeway" construction have been described in [168]. Unlike the "freeway" approach, OĎuńlaing et al. [159] and Yap [193] have developed a method of using *generalized Voronoi diagrams* for motion planning.

Gupta [75] and Gupta and Guo [76] have presented a backtracking motion-planning algorithm for manipulators with many degrees of freedom. The crux of their approach is to sequentially plan the motion of each link without explicitly representing and searching *high-dimensional* configuration space.

The work described in this book investigates the role of qualitative spatial reasoning as an approach to path planning. Such an approach utilizes a qualitative spatial characterization of robot free-space. When incorporated with a randomized local search technique (simulated annealing), it can generate near-optimal paths for robots with many degrees of freedom. In this approach, prior to planning an exact path, a general route that qualitatively satisfies certain optimality criteria is first constructed by searching a small number of disjoint regions in the free-space. The simulated-annealing search is then carried out only within the route (i.e., a sequence of connected regions) generated from the qualitative spatial planning. This approach reduces the computational time required in finding an exact path. More importantly, the qualitative approach is well suited for solving the problems where robot environments are not precisely known.

2.2.3 Local Path Planning

Several approaches have been proposed in the past that rely only on local information of the robot environment. Some of them are based on exact combinatorial algorithms, as described in [194], for planning the motion of a ladder or a polygon [131, 171]. Yap [195] has described an algorithm for planning exact local motions of a polygonal object through doors. Algorithms of direct practical application can be found in [57, 58].

In contrast to Yap's exact combinatorial algorithm, Chu and ElMaraghy [27] have proposed a *real-time* incremental path-planning approach, in which domain-specific heuristic rules are applied in synthesizing a collision-free trajectory. As has been shown in the mechanical assembly of a dish washer power unit, this approach is well suited for the real-time collision avoidance in a *multi-robot* task environment.

In other approaches, the configurations of the free-space are associated with a penalty function that characterizes, in general, the distance between the robot and its obstacles [89, 92, 105, 111]. The basic idea is to compute a repulsive potential field around each obstacle and an attractive potential field around the goal. The combination of both fields will force the robot to move away from the obstacles and toward the goal. However, the drawback associated with this approach is that local minima of the penalty function can sometimes lead to dead-ends during planning. This difficulty may be overcome either by incorporating global planning methods such as the $\mathcal{C}$-space method [140], by defining a harmonic penalty function that eliminate local minima [107], or by employing random motions at local minima [7].

2.3 Qualitative Reasoning

This section reviews previous work on the application of qualitative reasoning in the areas of mechanism analysis, spatial modeling, and robotics. It also highlights the earlier work on qualitative physics that has in part inspired the present study.

2.3.1 Qualitative Mechanism Analysis

Rather than being purely quantitative, recent studies on mechanism analysis have attempted to address the problem of integrating qualitative knowledge into the quantitative computations. Kramer [117] has reported a mechanism analysis system that can find the configurations of a set of rigid bodies satisfying geometric constraints by means of symbolic reasoning about degrees of freedom. This system utilizes a set of heuristics on how each constraint can be incrementally satisfied by a sequence of actions (e.g., translation or rotation).

The most significant work on *qualitative mechanism analysis* is that of Faltings [49, 50, 51]. In his work, Faltings has developed a first-principle algorithm for analyzing planar mechanisms, and has introduced a theory of *place vocabulary*. In his approach, the key to qualitative spatial reasoning is to compute distinct legal contact configurations from the quantitative $\mathcal{C}$-space (configuration-space) representation of a mechanism. The connectivity relationships are the primary constituents of qualitative states. The transitions of kinematic states during pairwise object interactions are derived via envisionments. In addition, Faltings has shown how possible kinematic topologies (e.g., connectedness of configuration space) may be derived directly from a symbolic description of the objects and qualitative information about their relative dimensions [53].

Dealing directly with real-life engineering problems, Pu [163] has demonstrated with an implemented program how the dynamic behavior of various *mechanical devices* can be qualitatively simulated. Faltings and Sun [55] have developed a formalism for modeling qualitative mechanical functions and a technique for generalizing a qualitative analysis of a novel device in order to extend domain knowledge.

Other studies on reasoning about mechanism behaviors have focused mainly on predicting mobility of mechanism parts and envisioning discrete dynamic or kinematic states [20, 63, 96, 160, 176].

Joskowicz and Sacks [98] have presented a composition algorithm for constructing *uniform motion regions* of mechanisms. The algorithm incrementally enumerates component sets to build motion regions, and is limited to fixed-axes, gear-like mechanisms. Kim [106] has described a qualitative motion analysis method for two-dimensional linkages where the variables take only three possible values, $\{-,\ 0,\ +\}$. The motion envisionments generated by Kim's method are too ambiguous to be practically useful.

In qualitative design of mechanisms, the problem of modifying objects' shapes to obtain certain motion transfer or constraint functions has been addressed by Faltings [52] and Joskowicz and Addanki [97]. In their approaches, the design of mechanism shapes is basically viewed as an inverse process of the qualitative kinematic analysis using place vocabularies. In doing so, a goal-directed causal analysis method is used to activate appropriate refinement operators for modifying shapes.

Kota [115] has proposed a qualitative matrix representation scheme for modeling the design building blocks of a mechanism. With qualitative matrices, the information on the interchangeability between input and output via a certain type of motion is captured. Such matrices allow one to generate possible structural forms of the mechanism. A slightly different schematic synthesis approach has been studied by Ulrich [184] and Ulrich and Seering [185], in which the modification of a candidate schematic description is conceived based on a minimal characterization (i.e., *compact representation*).

Nayak et al. [158] have presented a method for constructing device models by automatically selecting appropriate models for each of the device's components. The selection process depends on the contextual information of the components.

2.3.2 Qualitative Spatial Reasoning

Route Planning

Route planning is concerned mainly with finding the topology of a feasible path from one place to another. The earlier pioneering work on route planning has been done by Kuipers [119], Kuipers and Byun [122], Davis [31, 34], and McDermott and Davis [151]. In the *TOUR* program developed by Kuipers [119], the topological knowledge required for planning is represented as a set of *incidence relations* between places and roads. The program can assimilate and infer cognitive maps through the execution of `go_forward` and `turn` actions. To represent and reason about the metric details of the environment, McDermott and Davis [151] have proposed a spatial data structure called *fuzzy maps*, where the relevant object attributes, such as position, orientation, and height, are described in terms of *ranges*.

Spatial Inferencing

Mukerjee [156] argues that traditional quantitative geometric models are not suitable for abstracting the underlying spatial information needed for tasks such as planning. As an alternative representation scheme, he has introduced a set of qualitative spatial relations based on interval logic. Along the same line, Gusgen [77] has adopted Allen's qualitative temporal reasoning approach [2] to the spatial domain by aggregating multiple dimensions into a Cartesian framework. However, this approach fails to adequately capture the spatial interrelationships between individual coordinates.

Narayanan and Chandrasekaran [157] have proposed the idea of predicting spatial interactions between two-dimensional objects based on visual cases, or experience-based knowledge. In the context of path planning, Faltings and Pu [54] have explored a means-end approach in which path plans can be generated *incrementally*.

Randell [165] has developed a representation formalism for describing qualitative spatial relationships between concave objects using a list of qualitatively different primitive relations. In order to find a natural and efficient way for dealing with incomplete and *fuzzy* knowledge, Freksa [68] suggests a perception-based approach to qualitative spatial reasoning. Egenhofer and Franzosa [44] have developed a formal approach to describing spatial relations between point sets in terms of the intersections of their boundaries and interiors.

Frank [66] has discussed the use of orientation grids (cardinal directions) for spatial reasoning. A more detailed example of such an approach is given by Carney and Brown [22], where a framework for qualitative reasoning about shape and fit in the plug-and-socket domain has been built. Within this framework, the task of testing the fitness of two geometric surfaces is carried out in five stages: feature grouping, topological analysis, orientation, feature matching, and paired feature inspection.

Several researchers have investigated the computational representations of object shapes. Kimia [108] puts forward a complete theory of shape that unifies other approaches. His theory defines parts, protrusions, and bends as the basic elements of shape. Chang and Jungert [25] have developed a knowledge structure for describing relations between arbitrarily shaped two-dimensional objects on the basis of string representations. Lee and Hsu [130] have used a similar representation and developed a picture algebra for rectangles. Bard and Troccaz [6] have conducted a study in the domain of preshaping grippers. In their study, the process of preshaping a mechanical gripper during the task execution is guided by a *qualitative model* of the object to be manipulated, which can be automatically extracted from low-level visual data. The preliminary results have shown that this approach is promising.

Forbus et al. [63, 64] point out that no purely qualitative spatial reasoning method exists to adequately produce non-ambiguous results. Therefore, the goal of qualitative spatial reasoning is to provide approximate solutions which may be used to guide the application of quantitative methods. In this book, the results generated from the qualitative envisionments are incorporated into simulated-annealing-based quantitative mechanism analysis.

2.3.3 Qualitative Robotics

Another active research area, which is closely related to both mechanism analysis and spatial reasoning, is the application of qualitative models and reasoning techniques to directly solve the specific problems of *robotics*, such as robot task planning, motion planning, and robot vision.

Liu [132] has presented a framework of qualitative kinematics based on qualitative spatial inferencing techniques. The spatial inferences are generated based on a set of naive trigonometric rules. This approach can be applied to qualitative robot task planning [136, 134, 135]. A similar trigonometry-based approach has been proposed by Blackwell in the context of geometric scene description [12]. The work presented in this book further extends the previous qualitative trigonometric reasoning approach, and provides a *complete* representation and inference formalism.

2.3.4 Qualitative Physics

Several significant theories of qualitative reasoning have been introduced in the past [13, 14, 88]. Among them, *qualitative component-based analysis* [36, 37], *qualitative constraint-based simulation* [121], and *qualitative process-based modeling* [61] have been most widely applied. These theories provide many insights into the *general usefulness of qualitative reasoning* in modeling and simulating the behavior of a physical system.

Although the general intent of qualitative approaches is to capture the *deep incomplete knowledge* underlying human experts' reasoning in analyzing physical realities (e.g., systems functionality), there exist some methodological variations among different theories [15].

The *qualitative component-based analysis* formulated by deKleer and Brown [37] suggests that the behavior of a system may be determined by interrelating the behaviors of its components according to connectivity. In this case, the system is first decomposed conceptually into a collection of components and then the behavior of each component is modeled as a set of constraint confluences [28, 190].

Kuipers [120, 121] has developed a constraint-based approach to qualitative reasoning about physical behavior. In his well-known *qualitative simulation* theory, the constraint model of a system is directly derived from a set of observable parameters as well as their mathematical interrelations. In most cases, this approach yields similar results to deKleer and Brown's, however in a much more efficient way.

Forbus's *qualitative process theory* is more concerned with the active processes underlying physical realities [61]. It provides a tool for expressing and reasoning about more *intuitive* notions commonly used by humans [38, 187].

The above theories of qualitative reasoning have been widely applied to various engineering domains. One of the most significant applications is the work of Yoshikawa et al. [109, 179, 180]. Their work successfully demonstrates that for intelligent computer-aided design, the physical behavior of a design object can be adequately modeled using qualitative representations.

There have also been some extensions to the aforementioned qualitative reasoning theories. For example, Forbus [62] has further formulated a qualitative simulation of physical systems, which includes the effects of an agent's actions. Raiman [164] has extended the representation and reasoning in qualitative physics by considering the influence of different phenomena according to their relative *orders of magnitude* and by using this information to distinguish among radically different ways in which a physical system behaves. Simmons [175], having integrated qualitative and quantitative reasoning, and having combined inequality with simple addition and multiplication, has introduced *commonsense arithmetic reasoning.*

Mavrovouniotis and Stephanopoulos [149] have formalized order-of-magnitude reasoning with semantics, whereas Dague [30] has tackled the problem of expressing gradual changes in order of magnitude. Dubois and Prade [40] model the relative order-of-magnitude relations by domain-dependent fuzzy relations or semantics. Travé-Massuyès et al. [181] have proposed an axiomatic theory of qualitative equality and a general qualitative algebra which allows the use of any quantity space.

Most of the qualitative reasoning techniques, as mentioned above, require propagating qualitative values in a constraint network. With an emphasis on the label-based (e.g., signs or intervals) inferencing, Davis [32] has presented a comprehensive review of various constraint-propagation schemes.

De Mori and Prager [155] have further extended the interval-based qualitative calculus, and developed a method for analyzing the perturbations of linear dynamic system models. The qualitative perturbation analysis approach has been successfully applied to the domain of aerodynamic simulation, which results in a knowledge-based system (called the *Flite* system) to assist simulation engineers in validating flight model programs [162].

In qualitative physics research, apart from modeling and reasoning about deep physical knowledge, attention has also been paid to the development of formal theories to represent human *intuitive knowledge* of physical realities. Thus, instead of deriving functions from physical structures as mentioned above, the fundamental methodology used in this approach is to axiomatize commonsense by proposing topological relations, time ontologies, and primitives, etc. The representative work done in this area is Hayes's naive physics [82, 83, 84].

Davis [33] has developed a first-order logical framework for commonsense reasoning about the behavior of solid objects based on the principle of physics (without considering physical interactions in great detail). Kook and Novak [113] have used *canonical physical objects* and *physical models* to describe the behavior of physics problem solving. Shoham [173] reformulated some of the classic kinematics work developed in the last century and proposed a naive theory of the freedoms for planar objects. Mitchell et al. [153] have explored the application of naive physics to guide a robot hand-eye system in manipulating its world, and in acquiring refined knowledge about manipulation in specialized contexts. Other relevant work includes Funt's diagrammatic modeling of human perception of physical object interactions [69].

2.4 Simulated Annealing

Simulated annealing is the simulation of heat transfer (cooling) in a physical system in order to bring the system to a state where its energy is at a global minimum. Metropolis et al. [152] have developed a simulated-annealing algorithm that generates a sequence of states. The sequence is a Markov chain in which each state depends on the previous state according to the Boltzmann distribution. Hajek [78] has provided a review of the simulation algorithms with discussions on some of their variants.

Simulated annealing has been applied as an optimization technique to solve the *Traveling Salesman Problem*, a well-known NP-hard problem [23] [110].

Siarry et al. [174] have studied the problem of deriving optimal block placement in a VLSI design. Their work has investigated the effects of some parameters, such as *initial temperature*, on the performance of the optimization, and has explored the interactive use of the technique in order to reduce computation time.

Although simulated annealing has been used in solving various classical optimization problems, so far few complete analytical results exist concerning the performance of the technique. Hence, experimental studies are necessary in order to investigate the effects of annealing parameters. The key parameter in simulated annealing is the *annealing schedule*. A number of schedules have been proposed in the past, as summarized in [29].

3

Interesting Problems in Spatial Reasoning and Planning

This chapter provides formal statements of the problems which are the central focus of this book. It begins by defining the fundamental terminologies as well as their notations. In order to isolate the issues that are to be investigated in depth, some assumptions underlying the current work are proposed.

3.1 Terminology and Notation

In this book, the term *mechanism* refers to a collection of rigid bodies whose relative positions are constrained by their connecting joints. The individual parts of a mechanism, regardless of their shapes or the type of connections between them, are called *links*. The joints, which constrain the relative motions of individual links, are known as *kinematic pairs*.

Generally speaking, the kinematic pairs can be classified into two categories, *lower* and *higher pairs*. As defined by Reuleaux [166], lower pairs are those having surface contacts (e.g., prismatic joints), whereas higher pairs are those having line or point contacts (e.g., rolling contacts).

In the context of the current work, only linkage mechanisms with lower pairs are considered. In particular, the mechanisms whose links are connected by either revolute or prismatic joints are investigated. Such a mechanism can be represented as follows:

$$\mathcal{M} = \{L_1, L_2, ..., L_n; J_1, J_2, ..., J_k\} \quad (3.1)$$

which denotes an n-link k-joint planar mechanism. Here, each joint is assumed to be connected to only two links.

Depending on the connections of the links, a mechanism can also contain closed loops in its chain of links.

Definition 1 (Closed-chain and open-chain mechanisms). *A planar mechanism is called a closed-chain mechanism if it contains at least one closed loop in its chain of links. Otherwise, it is called an open-chain mechanism.*

Depending on the degrees of freedom (*dof*) allowed, it is possible to distinguish *constrained mechanisms* from *under-constrained mechanisms.*

Definition 2 (Constrained mechanisms). *A mechanism, composed of a set of moving and stationary rigid bodies connected by lower-pair (i.e., surface contact) joints, is called a constrained mechanism if its degree of freedom is exactly equal to one.*

A four-bar closed-chain linkage (or a four-bar linkage) is a typical example of such a mechanism (i.e., $dof = 1$).

Definition 3 (Under-constrained mechanisms). *A mechanism is called an under-constrained mechanism if its degree of freedom is more than one.*

Throughout the book, an under-constrained mechanisms is referred to only as a kinematic chain made of n rigid links $\{L_1, L_2, ..., L_n\}$ ($n \geq 1$). Any two consecutive links are connected by either a revolute joint (i.e., a hinge) or a prismatic joint (i.e., a sliding joint). An articulated robot may be considered as an example of such an under-constrained mechanism (i.e., $dof > 1$).

With respect to the constrained mechanisms, the terms of driver and follower links are sometimes used.

Definition 4 (Driver and follower links). *A driver link is the part of a mechanism which causes motion. A follower link is the part of a mechanism whose motion is affected by the motion of the driver.*

In the book, if not explicitly stated, the positions of a mechanism in a two-dimensional Euclidean space are always specified with respect to a *fixed frame of reference.* The notion of *relative motion* of a body refers to its movement relative to another body that may also be moving with respect to the fixed frame.

Definition 5 (Generalized and relative coordinates). *In kinematic analysis, if a mechanism has n degrees of freedom, and its configuration is described by only n coordinates, then these coordinates are referred to as the generalized coordinates. In a planar constrained mechanism, there will be only one generalized coordinate. On the other hand, if the selected coordinates define the position and orientation of each moving body with respect to a non-moving body or with respect to another moving body, then these coordinates are referred to as relative coordinates.*

In under-constrained mechanisms with only revolute joints, it is assumed that L_1 is the first link connected to a fixed frame of reference by a joint, and the configuration $\mathbf{C}$ is represented by a list of n parameters, $(\theta_1, \theta_2, ..., \theta_n)$, where $\theta_i \in \mathcal{R}$, $\mathcal{R} \subseteq \Re$, and $\Re$ is the set of all real numbers. As illustrated in Figure 3.1, the parameter θ_i is used to specify the position and orientation of L_i *relative* to L_{i-1}. θ_i corresponds to a revolute joint with no mechanical stop; that is, the range of θ_i is equal to $[0, 2\pi)$ (note: modulo 2π arithmetic is used).

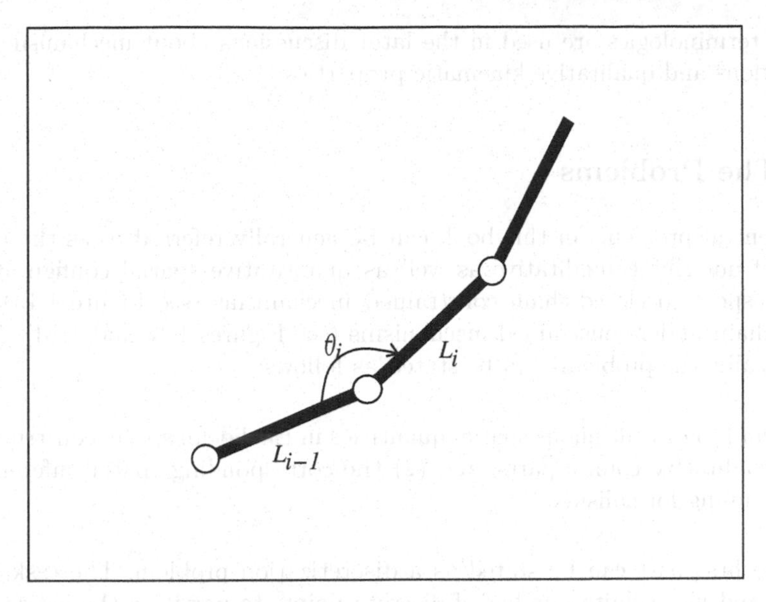

Fig. 3.1. In under-constrained mechanisms with only revolute joints, parameter θ_i specifies the position and orientation of L_i *relative* to L_{i-1}.

Definition 6 (Instantaneous linear and angular positions). *An instantaneous linear position ((x, y), $x, y \in \Re$), which is measured at a given instant, is defined as the two-dimensional Cartesian position of a moving point with respect to some reference frame. An instantaneous angular position ($\theta \in \Re$), which is the position found at a given instant, is defined as the angle between the position of a rotating point and a reference frame originated at its rotational axis.*

Definition 7 (Linear and angular displacements). *A linear displacement ($\Delta s = (\Delta x, \Delta y)$) is the change in linear position of a moving point during some time interval. An angular displacement ($\Delta\theta$) is the change in angular position (or the amount of rotation) of a rotating point during some time interval.*

Definition 8 (Velocity). *Velocity (V) is defined as the rate of change in position of a moving point with respect to time. Instantaneous velocity is the velocity found at a given instant. Relative velocity of a moving point A with respect to another point B, $V_{A/B}$, is defined as*

$$V_{A/B} = V_A - V_B \tag{3.2}$$

where V_A and V_B denote the velocities of points A and B relative to some reference frame, respectively.

This section has defined the fundamental kinematic concepts such as kinematic chain, constrained and under-constrained mechanisms, and positions.

These terminologies are used in the later discussions about mechanism configurations and qualitative kinematic properties.

3.2 The Problems

The central problems in this book can be generally referred to as the problems of describing qualitative as well as quantitative spatial configurations with respect to closed-chain constrained mechanisms (see Figure 1.1a) and open-chain under-constrained mechanisms (see Figures 1.1c and 1.1d). More specifically, the problems can be stated as follows:

Problem I. For continuous spatial quantities in Euclidean space, construct (1) their qualitative counterparts, and (2) the corresponding spatial inferencing and planning formalisms.

The first part can be stated as a discretization problem. The task is to define and use a finite number of discrete points to partition the entire continuous domain $\mathcal{R} \subseteq \Re$ of x into a finite set of mutually disjoint, bounded or unbounded subdomains, $\mathcal{Q} = \{Q_1, Q_2, ..., Q_m\}$, where $\bigcup_{i=1}^{m} Q_i = \mathcal{R}$. Each subdomain, such as an interval, is given a qualitative label. The spatial qualitative quantities of particular interest are Euclidean distance, linear displacement, angular displacement, and relative position and orientation.

The second part can be stated as the problem of symbolic reasoning based on the qualitative spatial relations. In spatial inferencing, the task is to construct basic *inference rules* for logically deducing spatial relationships. In spatial planning, the task is to apply *heuristic search* to find a sequence of qualitative spatial configurations from an initial configuration to a goal (final) configuration. The search criteria include not only satisfying existing qualitative spatial constraints but also satisfying certain optimality objectives.

A typical qualitative spatial inferencing problem is as follows:

> *Given three points a, b, and c in a plane where the Euclidean distance between a and b is* `much less` *than the distance between a and c, and the angle of the orientation* $\overline{ab}$ *(denoting the line segment joining a and b) with respect to* $\overline{ac}$ *is* `slightly acute`*, deduce the relative orientation of* $\overline{bc}$ *and* $\overline{ac}$*, and the distance between b and c.*

A typical qualitative spatial planning problem can be stated as follows:

> *Given a region-based qualitative representation of free space, find the shortest route from region* $\mathrm{N_0}$ *to region* $\mathrm{N_k}$*. The qualitative route is specified by a sequence of* $\{\mathrm{N_0}, \mathrm{A_0}, \mathrm{N_1}, \mathrm{A_1}, ..., \mathrm{A_{k-1}}, \mathrm{N_k}\}$*, where* $\mathrm{A_i}$ *denotes a transition from region* $\mathrm{N_i}$ *to region* $\mathrm{N_{i+1}}$*. The length of the route is defined as the number of transitions it requires, corresponding to the number of regions traveled.*

Problem II. Characterize motion paths as well as velocity relationships of a closed-chain constrained linkage mechanism.

The problem of constrained path analysis of a certain point in a mechanism can be translated into the following: First, the angular displacement of a driver link (as an input) is partitioned into *a finite set of ordered regions*, corresponding to a set of distinct qualitative configurations (or kinematic states). Then, for each input position, its qualitative instantaneous configuration is described, and accordingly its exact positions are located. Hence, for any point on a given link, a sequence of instantaneous positions of the mechanism can be derived with respect to a fixed reference frame. These positions constitute a discrete representation of the *constrained path* pertaining to a specific point.

The *instantaneous configuration* of a mechanism is determined by the instantaneous positions of individual links in the mechanism. The instantaneous configuration analysis problem can be stated as follows:

> *Given the qualitative spatial relationships between the lengths of links and the qualitative value of one generalized coordinate (i.e., a linkage with its dimensions and one of its joint angles being specified), find the positions of links in terms of their joint angles.*

In Chapter 6 of this book, the above problem will be treated in two general steps. The first step (the qualitative step) involves qualitative envisionments based on the qualitative spatial inferencing technique proposed for Problem I, whereas the second step (the quantitative step) applies a simulated-annealing technique to search for exact configuration positions based on the numerical ranges (or intervals) of the spatial variables as specified by the qualitative configurations inferred from the first step.

In the qualitative step, a closed-chain mechanism is represented as a collection of connected line segments such that its configurations can be described in terms of the lengths of line segments L_i and their joint angles θ_j.

Given the dimensions of the mechanism associated with a qualitative spatial relationship representation, the task of deriving velocity relationships is essentially a task of constructing an *instantaneous rotation center* which is configuration-dependent. Hence, this problem can be reduced to that of instantaneous configuration analysis.

Problem III. Find the path of an open-chain under-constrained mechanism geometrically constrained by other planar objects.

In this problem, two types of under-constrained mechanisms are considered, namely a *single mobile object* — a rigid body of convex polygon shape, whose position and orientation relative to a reference frame can be specified by $\mathbf{C}=(x, y, \theta)$ and an *open-chain linkage* — a kinematic chain made of n rigid links $\{L_1, L_2, ..., L_n\}$ with either revolute or prismatic joint connections. In general, an n-link open-chain linkage requires n parameters to describe its instantaneous configuration, i.e., $\mathbf{C}=(\theta_1, \theta_2, ..., \theta_n)$. The *configuration space* of an under-constrained mechanism is the space spanned by its independent variables.

The *environment* of such mechanisms is composed of a set of stationary non-intersecting convex-polygonal objects, $\{P_1, P_2, ..., P_i, ..., P_n\}$ (a *polygonal environment*), with their exact geometry being specified.

Like Problem II, Problem III is also handled first qualitatively and then quantitatively. In this case, the central focus is on *synthesizing a feasible path in a polygonal environment*, as opposed to analyzing a kinematically-constrained path without considering its environment. Such a difference in the task is also reflected in the way in which the qualitative and quantitative steps are performed.

The qualitative phase applies the qualitative spatial planning technique developed for Problem I to find a route (a sequence of connected regions) in which an exact path may exist. Based on the results of the first step, the quantitative phase performs local path search using the same simulated-annealing technique as in Problem II with a different optimality (i.e., distance from the final goal).

Here, the *exact path* is defined as a sequence of instantaneous configurations of the mechanism connecting initial and goal configurations. In a region-based qualitative partition of Euclidean free-space, the path can be decomposed into three path segments, namely:

1. a starting path segment from $\mathbf{C}_{\text{init}}$ to $\mathbf{C}_{\text{m}}$,
2. a sequence of connected path segments from one $\mathbf{C}_{\text{m}}$ to another, and
3. an ending path segment from $\mathbf{C}_{\text{m}}$ to $\mathbf{C}_{\text{goal}}$.

where $\mathbf{C}_{\text{init}}$ and $\mathbf{C}_{\text{goal}}$ are the *initial* and *goal* configurations of the mechanism, respectively, and $\mathbf{C}_{\text{m}}$ is the configuration of the mechanism at the boundary of a qualitative region.

3.3 Assumptions

This section states the assumptions underlying the current work. These assumptions are adopted, primarily, for the purposes of demonstration and in-depth investigation of the above-mentioned problems. The assumptions are as follows:

1. In the case of an under-constrained mechanism (for example, see Figures 1.1c and 1.1d), the mechanism is the only moving object in the environment (i.e., the environment is *static*).
2. There is no direct contact between the mechanism and its environment during the motion (i.e., the mechanical interaction does not exist).

3. There is no closed loop in an under-constrained mechanism; in other words, the under-constrained mechanism will be either a single mobile object or an articulated open kinematic chain.
4. In the case of under-constrained mechanisms, the objects in the environment as well as the single mobile object are all of convex polygonal shape.
5. The joints of the planar mechanisms consist of only revolute joints (i.e., hinges) and prismatic joints (i.e., sliding joints).

4

How to Represent Qualitative Spatial Relationships

This chapter presents a formalism of *qualitative* measurement representation used to describe the spatial relations of planar objects. The qualitative spatial quantities (variables) of particular interest are Euclidean distance, linear displacement, angular displacement, relative position and orientation, and location. The notation $[x]$ represents a qualitative variable corresponding to a quantitative variable x.

Definition 9 (Qualitative quantity space). *Let x be a quantitative variable, such that $x \in \mathcal{R}$, and $\mathcal{R} \subseteq \Re$. If the entire domain $\mathcal{R}$ is partitioned into a finite set of mutually disjoint subdomains $\{Q_1, Q_2, ..., Q_m\}$, i.e., $\bigcup_{i=1}^{m} Q_i = \mathcal{R}$, and, furthermore, all numerical values lying within Q_i are treated as being equivalent and named symbolically by $Label(Q_i)$, then the qualitative variable $[x]$ corresponding to x is defined as follows:*

$$[x] \in X, \; X \subseteq \bigcup_{i=1}^{m} Label(Q_i) \tag{4.1}$$

where $Label(Q_i)$ is called a primitive qualitative value. The union of X is called the qualitative quantity space of $[x]$, denoted by $\mathcal{Q}\text{-}space_{[x]}$.

The two most fundamental spatial quantities in Euclidean space are *distance* and *angle*. The primary goal of formulating qualitative measurement spaces for distances and angles is to provide a general vocabulary for performing efficient symbolic spatial inferences and analysis. Although the semantics of the vocabulary may not be unique, the discrete values for its definition over the continuous domains should allow:

1. computationally *efficient* spatial inference (i.e., with a small number of labels) and
2. *less ambiguous* spatial descriptions (i.e., with sufficient precision).

In one of the most recent studies on qualitative spatial reasoning, Latecki and Röhrig [125] have proposed a technique for inferring qualitative angu-

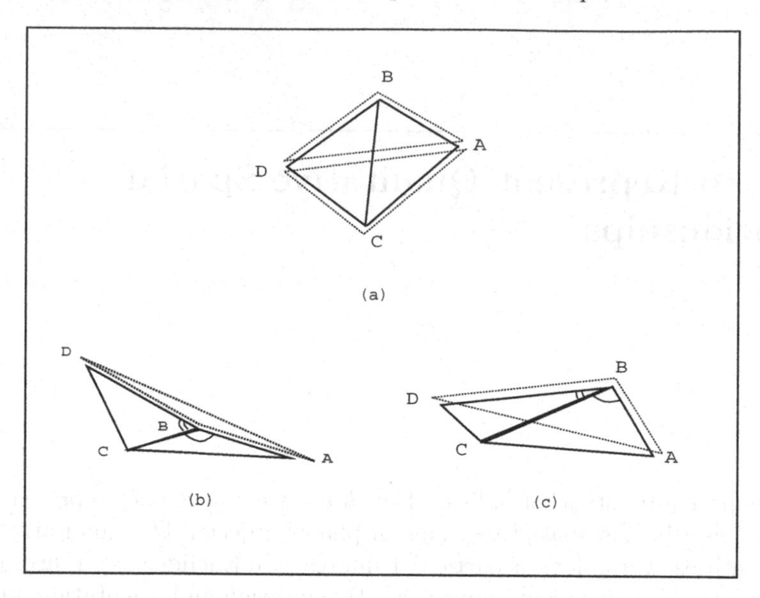

Fig. 4.1. Illustrations of Latecki and Röhrig's recent work [125]. (a) Given $\triangle ABC$ and $\triangle CBD$, qualitative descriptions of $\triangle ABD$ and $\triangle ACD$ are desired. (b) If two `counterclockwise` oriented angles, an `obtuse` $\angle ABC$ and an `acute` $\angle CBD$ are given, then it can be inferred that $\angle ABD$ is `obtuse`. (c) $\angle ABD$ is `obtuse`, but its orientation is different from that shown in (b).

lar relationships in a cognitive map. In their study, as illustrated in Figure 4.1a, the major effort is to show how triangles $\triangle ABD$ and $\triangle ACD$ can be qualitatively described given the angular orientations (i.e., `clockwise` or `counterclockwise`) of triangles $\triangle ABC$ and $\triangle CBD$ and the corresponding qualitative angles in terms of `acute` and `obtuse`. For example, if two `counterclockwise` oriented angles, an `obtuse` $\angle ABC$ and an `acute` $\angle CBD$, are given, it can be inferred that $\angle ABD$ is `obtuse`, as shown in Figure 4.1b. However, using this technique (with only four labels), the orientation of $\angle ABD$ cannot be inferred (note that the orientations of $\angle ABD$ can be either `clockwise`, as in Figures 4.1b, or `counterclockwise`, as in Figures 4.1c). Furthermore, if $\angle ABC$ is also `acute`, then $\angle ABD$ becomes completely uncertain. This ambiguity in the spatial descriptions may prevent the technique from practical application.

The following two sections introduce new measurement space for *qualitative distance* and *qualitative angle.*

4.1 Qualitative Distance

In qualitative spatial representation, the notion of a distance, say x, is always expressed in a relative sense. To perform spatial reasoning, it is useful to express a distance x relative to another reference distance d_{const}. A typical example of such an expression is "x is `much shorter` than d_{const}". The relation `much shorter` can be regarded as the qualitative value of a Euclidean distance relative to the reference constant. Formally, the variable *qualitative distance* is denoted by $[x \mid d_{\text{const}}]$. The *qualitative length* of a rigid link is defined, in the same way as the qualitative distance, with respect to the length of another link.

Another qualitative quantity which shares the same definition is that of linear displacement. The qualitative magnitude of a linear displacement from point p to point q is measured by the qualitative Euclidean distance from p to q, as denoted by $[\overline{pq} \mid d_{\text{const}}]$.

The $\mathcal{Q}$-space of a qualitative distance $[x \mid d_{\text{const}}]$ is determined by a set of qualitative labels defined over the numerical domain of the distance ratio x/d_{const}. In the current work, three distinct quantitative values, {2/3, 1, 3/2}, are used to partition such a numerical domain. Accordingly, five labels can be derived. This qualitative labeling has been inspired in part by the existing empirical results on human spatial cognition (see Section 4.3) and by commonsense observations. However, it should be mentioned that the resulting *label definitions* neither claim any psychological and physiological foundations nor exclude other possible partitions of the numerical domain. In fact, as will be shown in the later discussion (in Section 5.1.1), the qualitative spatial inferencing technique proposed in this book can be adapted to different label definitions.

Definition 10 (Qualitative distance labels). *The primitive values (labels) of a qualitative distance, $[x \mid d_{\text{const}}]$, are defined with respect to a constant distance d_{const}, as follows (see Figure 4.2):*

$$\texttt{Less} \stackrel{\text{def}}{=} \{x \mid x \in \mathcal{R}, x/d_{\text{const}} \in (0, 2/3)\} \tag{4.2}$$

$$\texttt{SlightlyLess} \stackrel{\text{def}}{=} \{x \mid x \in \mathcal{R}, x/d_{\text{const}} \in [2/3, 1)\} \tag{4.3}$$

$$\texttt{Equal} \stackrel{\text{def}}{=} \{x \mid x \in \mathcal{R}, x/d_{\text{const}} = 1\} \tag{4.4}$$

$$\texttt{SlightlyGreater} \stackrel{\text{def}}{=} \{x \mid x \in \mathcal{R}, x/d_{\text{const}} \in (1, 3/2]\} \tag{4.5}$$

$$\texttt{Greater} \stackrel{\text{def}}{=} \{x \mid x \in \mathcal{R}, x/d_{\text{const}} \in (3/2, \infty)\} \tag{4.6}$$

The semantics of `Equal` is straightforward, i.e., the distance x is equal to d_{const}. The semantics of `SlightlyGreater` is that x is not longer than one and a half of d_{const}, whereas `Greater` means that x is at least one and a half of d_{const}. `SlightlyLess` and `Less` are defined as the inverse values of `SlightlyGreater` and `Greater`, respectively. Hence, given two distance variables A and B, the following equivalences are valid:

$$[A \mid B] = \texttt{Less} \Longleftrightarrow [B \mid A] = \texttt{Greater} \tag{4.7}$$

$$[A \mid B] = \texttt{SlightlyLess} \Longleftrightarrow [B \mid A] = \texttt{SlightlyGreater} \tag{4.8}$$

The labels `Less`, `SlightlyLess`, `Equal`, `SlightlyGreater`, and `Greater` are also denoted in the book as *'l', 'sl', 'eq', 'sg', and 'g'*, respectively.

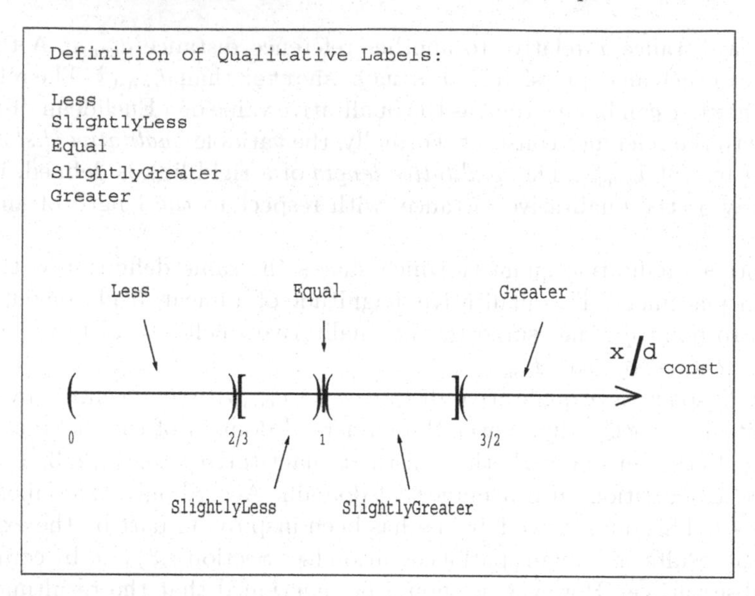

Fig. 4.2. Qualitative labeling of a relative-distance domain. The qualitative abstraction of a Euclidean distance, x, is obtained by partitioning the numerical domain of a ratio, x/d_{const}, into a set of disjoint intervals, where d_{const} is a reference distance. It should be noted that the choice of partitioning points can be domain-dependent and hence not unique.

4.2 Qualitative Angle

Suppose that starting from one point, O, there are two rays coming out. In the following presentation, their smallest angle is referred to as the *angle* between the two rays, denoted by θ. This angle always lies within the domain of $\mathcal{R} = [0, \pi]$.

To construct a $\mathcal{Q}$-space for $[\theta]$, a set of distinct quantitative values should be employed to partition the numerical domain $\mathcal{R}$. One possible partition is to use the values of $\{\pi/3,\ \pi/2,\ 2\pi/3\}$, as shown in Figure 4.3. This leads to five subdivisions.

Definition 11 (Qualitative angle labels). *The primitive values (labels) of a qualitative angle,* $[\theta]$, *are defined as follows:*

$$\texttt{Acute} \stackrel{\text{def}}{=} \{\theta \mid \theta \in [0, \pi/3)\} \tag{4.9}$$

$$\texttt{SlightlyAcute} \stackrel{\text{def}}{=} \{\theta \mid \theta \in [\pi/3, \pi/2)\} \tag{4.10}$$

$$\texttt{RightAngle} \stackrel{\text{def}}{=} \{\theta \mid \theta \ = \ \pi/2\} \tag{4.11}$$

$$\texttt{SlightlyObtuse} \stackrel{\text{def}}{=} \{\theta \mid \theta \in (\pi/2, 2\pi/3]\} \tag{4.12}$$

$$\texttt{Obtuse} \stackrel{\text{def}}{=} \{\theta \mid \theta \in (2\pi/3, \pi]\} \tag{4.13}$$

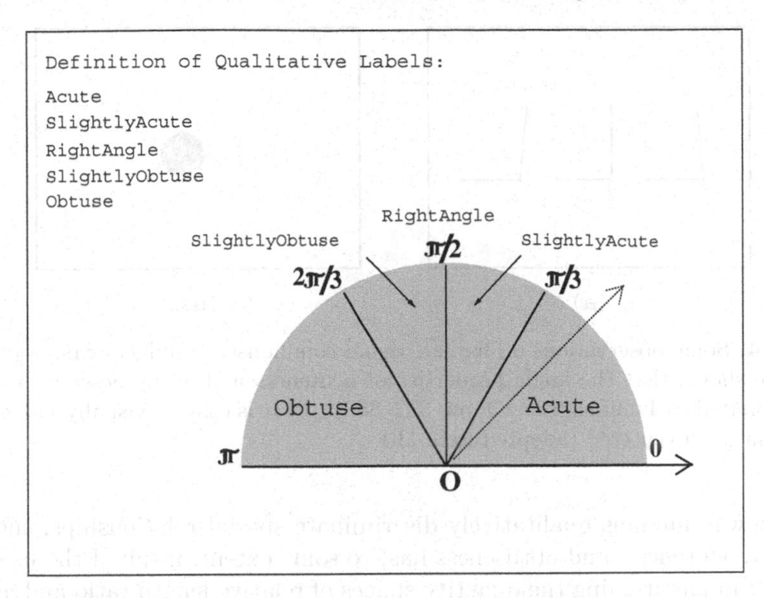

Fig. 4.3. Qualitative labeling of an angle domain. The semantics of qualitative angles are defined with respect to a set of disjoint numerical intervals. The exact partitioning points chosen for the intervals, although they should be consistent with human commonsense, may not be unique.

The labels of `Acute`, `SlightlyAcute`, `RightAngle`, `SlightlyObtuse`, and `Obtuse` are also denoted in the book as *'a', 'sa', 'r', 'so', and 'o'*, respectively.

4.3 Notes on *Label-Based Distance and Angle Descriptions*

The label definitions for qualitative distance and angle variables as presented in the preceding sections have been, to some extent, inspired by previous empirical findings on human commonsense visual judgements. As can be readily noted, humans are very good at making qualitative measurements with respect to some *symmetric* or *neutral* references.

Rock [167] (see p. 24) argues that human perception of size is in general characterized in relative terms (by contextual effects) such as relative lengths. More importantly, studies in psychology [3, 65, 99] have found that humans are proficient in visually judging the equality (or inequality) in the relative size of two objects or in discriminating an acute angle from an obtuse angle. For instance, as shown in Figure 4.4a, it is not difficult to visually judge that $\angle 1$ is acute, and $\angle 3$ is obtuse. Another example is given in Figure 4.4b, where there is no need for any computation in order to judge that the black point is off-centered with respect to the rectangular frame.

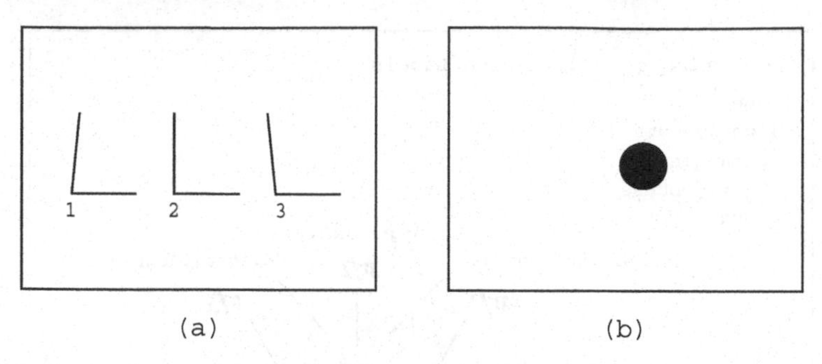

Fig. 4.4. Some observations on human visual commonsense judgements. (a) Studies have shown that the angle properties of acuteness and obtuseness can well be discriminated by humans (see [65] pp. 332–335). (b) It is easy to visually judge that the point is off-centered (adopted from [3]).

The way humans qualitatively discriminate spatial relationships, such as equality, acuteness, and obtuseness has, to some extent, inspired the use of 1 and $\pi/2$ in partitioning the quantity spaces of relative length ratio and angle, respectively.

As a result, two morphologies can be obtained; each has three elements (i.e., $\{<1,\ =1,\ >1\}$ and $\{<\pi/2,\ =\pi/2,\ >\pi/2\}$). However, such a level of precision will be insufficient to describe a variety of spatial relationships. Taking a simple triangle for example, each angle may be acute *to a certain extent*, and also, one side may be *slightly* longer than another. In order to cover as many geometric configurations as possible, and at the same time to avoid combinatorial explosion (caused by fine-grained partitions), 3/2 and its inverse 2/3 may be added to further divide the domain of the length ratio, and $\pi/3$ and $2\pi/3$ may be added to further divide the angle domain.

4.4 Completeness

Definition 12 (Completeness of a qualitative description). *A qualitative representation* $[x]$ *of* x *is complete if*

$$\forall x \in \mathcal{R},\quad \exists [x] \in \bigcup_i Label(Q_i) \subseteq \mathcal{Q} - space_{[\mathrm{x}]},\quad s.t.\ x \in \bigcup_i Q_i \tag{4.14}$$

where $\mathcal{R} \subseteq \Re$ *is the valid domain of* x. $Label(Q_i)$ *is the symbolic label of* Q_i, *which belongs to one of the primitive qualitative values.* $\mathcal{Q}$-$space_{[\mathrm{x}]}$ *is the qualitative quantity space of* $[x]$.

A *reasoning procedure* is said to be *complete* if and only if it yields complete qualitative descriptions.

A similar notion of completeness has been introduced by Kuipers [121] in his work on *qualitative simulation*. Kuipers has shown that the envisioning procedure, as used in the well-known qualitative simulation program *QSIM*, is complete.

4.5 Minimum-Spanning Edge (m-Edge) between Two Polygons

Definition 13 (Minimum-spanning line segments (edges)). *Let* $p = \{p_1, p_2, ..., p_n\}$ *and* $q = \{q_1, q_2, ..., q_m\}$ *be two non-intersecting convex polygons whose vertices are specified by their Cartesian coordinates in a clockwise order. The minimum-spanning line segment of* p *and* q *is defined as the line segment connecting the two polygons that has the shortest Euclidean distance (see Figure 4.5).*

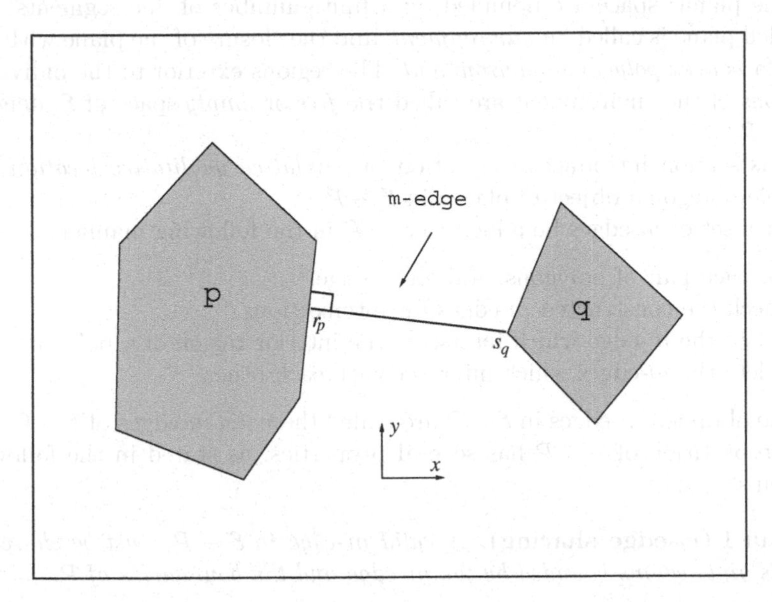

Fig. 4.5. The minimum-distance (also called minimum-spanning) line segment between the boundaries of two non-overlapping polygons is defined as the m-edge of the two polygons (see Eq. 4.15).

In the foregoing presentation, this line segment will also be called the *minimum-spanning edge*, denoted by m-edge$_{pq}$. Let L be a line intercepted by r_p and s_q which lie on p and q, respectively. The distance of m-edge$_{pq}$ satisfies the following condition,

$$d_{\text{m-edge}_{pq}} = \min_{\forall r_p \in p,\ \forall s_q \in q} \{d(r_p, s_q)\} \tag{4.15}$$

where $d(r_p, s_q)$ denotes the quantitative distance between points r_p and s_q.

Given a two-dimensional Cartesian coordinate reference frame, the orientation of one polygon with respect to another is specified by the smallest positive joint angle θ formed by the x-axis and the m-edge.

Based on the above construction of an m-edge within a certain reference frame, the *qualitative distance* and *qualitative orientation* of any two polygons can be expressed by computing $[d_{\text{m-edge}} \mid d_{\text{const}}]$ and $[\theta]$, where d_{const} is a comparative Euclidean distance of interest.

4.6 Qualitative Location in a Convex Polygonal Environment

Let $\mathcal{P} = \{P_1, P_2, ..., P_k\}$ be a set of k non-intersecting convex polygons fixed in a two-dimensional Euclidean space, $\mathcal{E}$. It is assumed that both the polygons and the planar space are bounded by a finite number of line segments. The bounded plane is called an *environment*, and the closure of the plane with $\mathcal{P}$ is called a *convex polygonal environment*. The regions exterior to the individual polygons of the environment are called the *free* or *empty* space of $\mathcal{E}$, denoted as $\mathcal{E} - \mathcal{P}$.

This section introduces the notion of a *relative qualitative location* of a movable polygonal object O placed in $\mathcal{E} - \mathcal{P}$.

Let a set of m-edges be added to $\mathcal{E} - \mathcal{P}$ in the following manner:

1. For each pair of polygons, add an m-edge.
2. Check the constructed m-edges for intersections.
3. Delete the m-edge which intersects the interior region of a polygon.
4. Delete the m-edges which intersect with each other.

The obtained m-edges in $\mathcal{E} - \mathcal{P}$ are called the *valid* m-edges of $\mathcal{E} - \mathcal{P}$. The m-edge partition of $\mathcal{E} - \mathcal{P}$ has several properties, as stated in the following theorems.

Lemma 1 (m-edge sharing). *A valid m-edge in $\mathcal{E} - \mathcal{P}$ must be shared by two disjoint regions bounded by the m-edge and the boundaries of $\mathcal{P}$.*

Proof: Suppose there exists an m-edge$_{\text{i}}$ which does not divide two regions. Then it is possible to construct a polygonal chain starting from one side of the m-edge$_{\text{i}}$, going through a single region, and ending at the opposite side of m-edge$_{\text{i}}$ without intersecting other m-edges (note that if it intersects with other m-edges, the lemma is proven due to the fact that the number of bounding m-edges for the region is finite). From the polygon(s) enclosed by such a polygonal chain, construct m-edges with other non-enclosed polygons, as shown in Figure 4.6. Whenever the constructed m-edge intersects with an existing m-edge$_{\text{x}}$, build an m-edge to the polygons that m-edge$_{\text{x}}$ connects to, and repeat the above procedure recursively. According to the hypothesis, for each new m-edge there must exist a new intersecting m-edge$_{\text{x}}$. Each time,

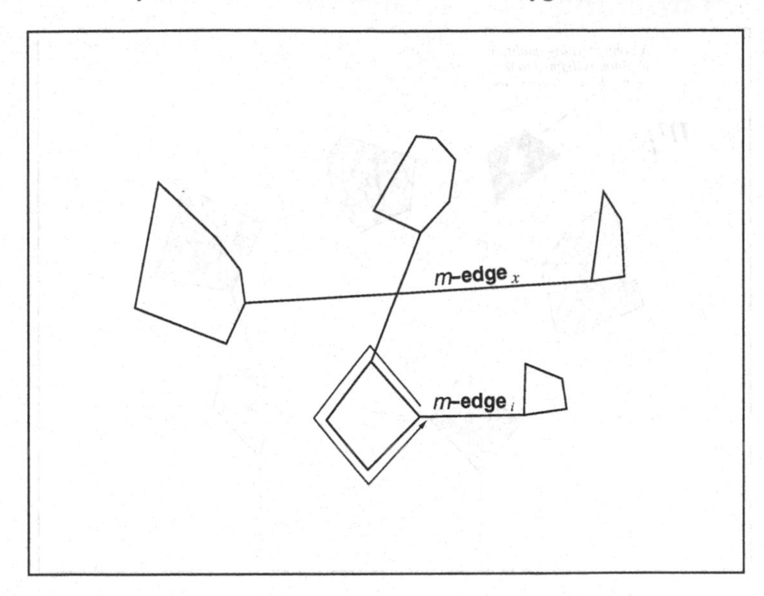

Fig. 4.6. The construction of m-edges from an isolated set of polygons – the proof of Lemma 1.

m-edge$_x$ connects to two different polygons. This implies that the number of polygons is infinite, a contradiction. ∎

Theorem 1 (Mutually disjoint partition). *The m-edge partition of $\mathcal{E} - \mathcal{P}$ will result in a set of mutually disjoint subdivisions.*

The proof of Theorem 1 immediately follows from Lemma 1.

Definition 14 (m-closure regions). *The covering set of disjoint subdivisions $\{Q_1, Q_2, ..., Q_m\}$ formed by the boundaries of $\mathcal{P}$ and the valid m-edges of $\mathcal{E} - \mathcal{P}$ is called a set of m-closure regions, as shown in Figure 4.7. The partition of $\mathcal{E} - \mathcal{P}$ into m-closure regions is also called an m-closure partition.*

Theorem 2 (Uniqueness). *The m-closure partition of $\mathcal{E} - \mathcal{P}$ is unique.*

Proof: Suppose that there exist two different m-closure partitions. Based on the m-edge partition procedure stated earlier, it is known that there must be some m-edge removed in the first partition but not in the second, or vice versa. This implies that the partitions must be constructed in two different environments, a contradiction. ∎

Theorem 3 (Non-convex polygonal environments). *For non-convex polygonal environments, $\mathcal{E} - \mathcal{P}$ cannot always be subdivided into a set of mutually disjoint m-closure regions.*

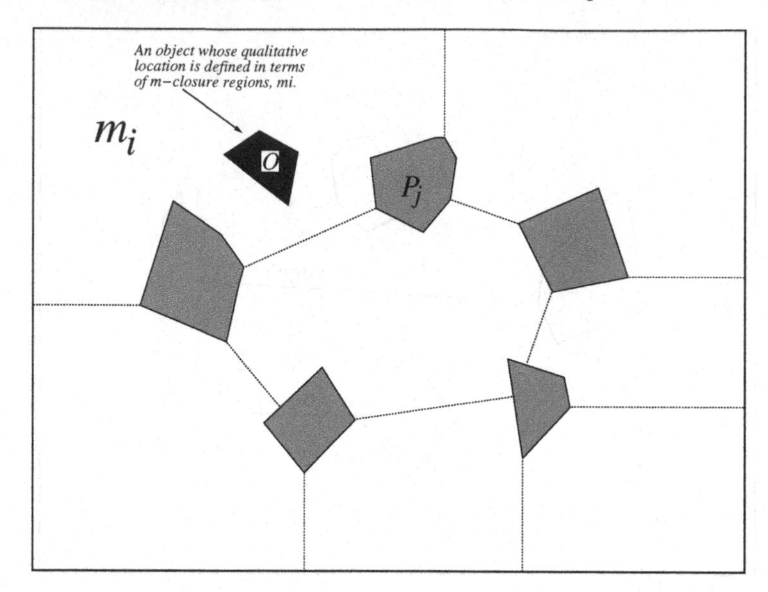

Fig. 4.7. The m-closure partition of a polygonal environment. The set of m-closure regions (m_i) is obtained by constructing *non-intersecting* m-edges between polygons (P_j).

Proof: See Figure 4.8. ■

One way to handle such cases is to transform a non-convex polygon into a set of disjoint convex polygons and then partition the transformed convex polygonal environment.

4.6.1 Qualitative Location

Definition 15 (Qualitative Location of an object). *The qualitative location of a planar object O, denoted as $[LOC_o]$, is specified by a set of adjacent m-closure regions with which O overlaps, as shown in Figure 4.7:*

$$[LOC_o] = \{m_i \mid m_i \in \mathcal{E} - \mathcal{P},\ O \cap m_i \neq \emptyset\} \tag{4.16}$$

where m_i denotes an m-closure region, and $\emptyset$ is an empty set.

The qualitative locations of two points x and y are said to be equivalent if they are both disjoint from $\overline{m_i}$ (i.e., both lie inside the same m-closure region, m_i).

By definition, an m-closure region m_i is closed by m-edges constructed between a set polygons $\{P'_1, P'_2, ..., P'_j\} \subseteq \mathcal{P}$. Hence, each m-closure region m_i in the above definition of a location also implies the qualitative orientations of O with respect to polygons $\{P'_1, P'_2, ..., P'_j\}$.

Fig. 4.8. An example of the non-convex polygonal environment which cannot be subdivided into a disjoint set of mutually disjoint m-closure regions – the proof of Theorem 3.

4.7 Graphic Representation of the m-Edge Partitioned Free-Space

This section describes a graphic representation for the m-closure partition of $\mathcal{E} - \mathcal{P}$.

Given a two-dimensional free-space $\mathcal{E} - \mathcal{P}$, its m-edge partition can be represented by a finite connected graph where nodes are m-closure regions, as shown in Figure 4.9, and arcs indicate the existences of adjacency relationships, i.e., m-edges shared by two distinct regions. Formally, such a *connectivity graph* can be denoted by an ordered quadruple $< N, A, c, t >$ where:

- N: a finite set of nodes.
- A: a finite set of arcs.
- c: a qualitative function associated with each arc A, which gives the relative distance and orientation between two m-edge spanned polygons.
- t: a transition function associated with each arc A, which indicates an unordered pair of nodes called endpoints of A.

Figure 4.10 provides an example of the connectivity graph.

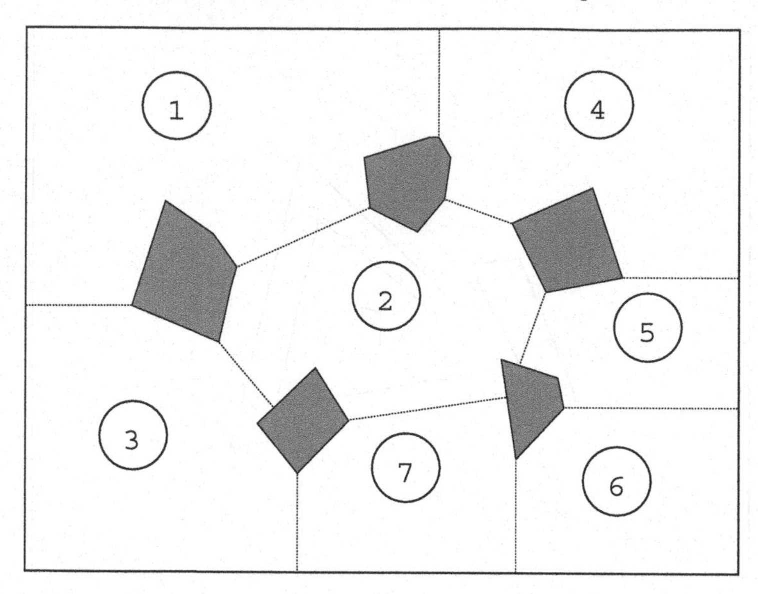

Fig. 4.9. In the free-space as shown in Figure 4.7, each of the identified m-closure regions can be represented as a graphic node.

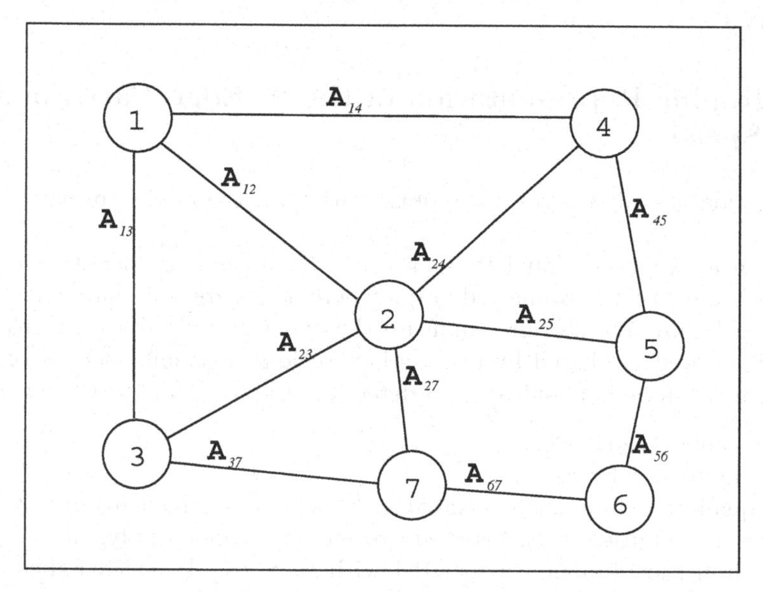

Fig. 4.10. A graphic representation of the free-space as given in Figure 4.7. Link A_{ij} denotes the connectivity between two nodes (regions) i and j.

4.8 Notes on *Qualitative Location*

In order to qualitatively describe the position as well as orientation of one object relative to a certain environment (in this section, it is assumed to be a polygonal environment), the notion of *qualitative location* has been formally defined. Kuipers [119] has previously introduced a similar notion, called places, to represent the topology of a robot environment. In his work on spatial map-building, the topological model of the environment is composed of a network of nodes and arcs, where nodes correspond to the *distinctively recognizable places* in the environment, and arcs correspond to path segments connecting the places, as illustrated in Figure 4.11.

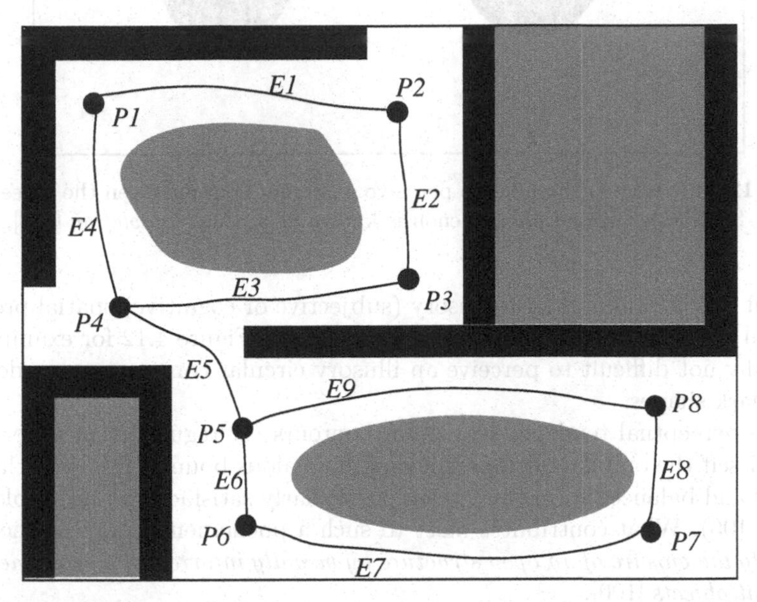

Fig. 4.11. The topological model of a robot environment, as used for cognitive map assimilation and inference in Kuipers's *TOUR* system [119]. In the diagram, $P1 \sim P8$ represent distinctive places, and $E1 \sim E9$ represent the path segments connecting the places.

By definition, a qualitative location is designated by m-edge closed region(s). This definition of qualitative location has been, in part, inspired by the way in which humans perceive and visually reason about two-dimensional figures (e.g., spatial maps).

As is well-known, what humans perceive may be enriched by "mental contents" not given in the external stimulus [167] (see p. 26). One of the phenomena in perception is known as *amodal completion*, which refers to perceptual existence that is not verified by any sensory modality [99, 100]. As Kanizsa and Gerbino [100] have observed, there are many cases where humans can,

Fig. 4.12. It is easy for humans to perceive a circular contour given the three dark regions. Such a perceptual phenomenon is known as *amodal completion* [100].

without much thinking, form illusory (subjective or cognitive) spatial properties and relationships. Taking the configuration of Figure 4.12 for example, it is usually not difficult to perceive an illusory circular contour surrounded by three dark shapes.

The perceptual tendency to regular contours (or regularity in general) is not in itself the only factor for achieving anomalous boundaries, even though regular and balanced forms may seem particularly satisfactory and stable [99] (see p. 106). What contributes most to such a phenomenon may be the *tendency to the closure of an open structure by visually interrelating or connecting adjacent objects* [100].

The above general observation implies that if an open area in a spatial map is surrounded by a set of closely spaced objects, then this area may be perceived as a closed region. This may be attributed to the human *intuitive spatial representation and planning* capability, whether based on visual images or based on internal cognitive maps. Figure 4.13 provides four images from which some local regions may be perceptually identified.

In this chapter, the representation of a polygonal environment based on m-edge partition is in fact designed to model the free-space with disjoint closed regions. These regions serve as an intermediate representation in the qualitative spatial planning, as will be discussed in Chapters 5 and 8.

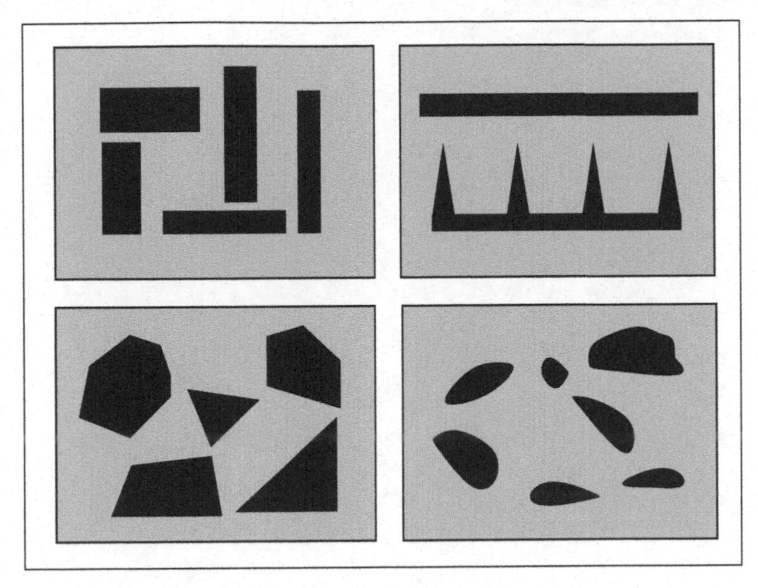

Fig. 4.13. Amodal completion may lead to the visual identification of regions in the images.

5

Methodology of Spatial Reasoning and Planning

Given a set of relative qualitative spatial relationships represented using the formalisms of Chapter 4 (e.g., the length and orientation of a link), two problems are of particular interest. First, how can one *infer* other new spatial relationships? Secondly, how can one *synthesize* a sequence of spatial configurations from an initial state to a goal state for a planar object which not only comply with the existing known geometric constraints but also satisfy some *optimality* criteria?

This chapter deals with the above-mentioned two problems; the former is referred to as *spatial inferencing* whereas the latter is referred to as *spatial planning*. Both spatial inferencing and planning require a *qualitative-envisionment* step for deriving possible qualitative spatial configurations. In spatial inferencing, the envisionment is created by propagating qualitative spatial relationships with the inference rules of *qualitative trigonometry* and *qualitative arithmetic*, whereas in spatial planning, the envisionment is created by using a heuristic search technique. The results of qualitative spatial reasoning will be used to guide a randomized local search (i.e., simulated annealing) in finding exact quantitative configurations.

Taking *spatial path planning* as an example, a global *route* that qualitatively satisfies certain optimality criteria is first constructed, prior to finding an exact path, by searching a small number of disjoint regions in the free-space. The simulated-annealing search is then carried out only within the route (i.e., a sequence of connected regions) generated from the qualitative spatial planning. This approach reduces the computational time required in finding an exact path. More importantly, the qualitative approach is well suited for solving the problems where complete geometric information is not available.

5.1 Spatial Inferencing

Spatial inferencing is concerned with the qualitative spatial analysis of the relative distances and relative orientations in two-dimensional Euclidean space. A typical spatial inferencing problem can be expressed as follows:

> *Given three points a, b, and c in a plane where the Euclidean distance between a and b is* `Less` *than the one between a and c, and the angle of the orientation $\overline{ab}$ (denoting the line segment joining a and b) with respect to $\overline{ac}$ is* `SlightlyAcute`, *deduce the relative orientation of $\overline{bc}$ and $\overline{ac}$, and the distance between b and c.*

The following sections describe the necessary spatial inference rules that will allow one to *logically derive* solutions to the above problem.

5.1.1 Qualitative Trigonometry ($\mathcal{QT}$)

Qualitative trigonometry ($\mathcal{QT}$) is concerned with the side-angle relationships in a planar triangle whose measurements are defined in the qualitative quantity space, as presented in Sections 4.1 and 4.2.

A preliminary version of $\mathcal{QT}$ was first introduced by Liu [132] as a set of inference rules to reason about two-dimensional spatial relationships. It was called *naive trigonometry* and dealt only with 18 geometrically-distinct triangles. The current work extends the previous formalism so that it can take into account broader triangular variations. Furthermore, the formulation presented here ensures that the derived $\mathcal{QT}$ rules are semantically *complete* (in the sense defined in Section 4.4).

Let three angles of a triangle be denoted by $\angle a$, $\angle b$, and $\angle c$, which are opposite to sides A, B, and C, respectively. Table 5.1 presents a tabular matrix of the qualitative spatial relationships of the planar triangle where the left-most column gives the qualitative lengths of A and B (in pairs) with respect to a certain reference length l_{ref} of interest, and the top-most row gives the qualitative joint angles $\angle c$ between A and B.

Each entry in the table specifies the corresponding possible length C and possible angle $\angle a$. As an example, the upper leftmost entry can be stated as follows:

> *Given that A and B are both* `Less` *than l_{ref}, and that angle $\angle c$ is* `Acute`, *it can be inferred that C is also* `Less` *than l_{ref} and $\angle a$ is* `Obtuse` *or* `SlightlyObtuse`.

Using the qualitative quantity notations, this inference rule can be translated further into the following expression:

$$([A \mid l_{\text{ref}}] = \texttt{Less}) \wedge ([B \mid l_{\text{ref}}] = \texttt{Less}) \wedge ([\angle c] = \texttt{Acute})$$

$$\Longrightarrow$$

$$([C \mid l_{\text{ref}}] = \texttt{Less}) \wedge ([\angle a] = \texttt{SlightlyObtuse} \vee \texttt{Obtuse})$$

Table 5.1. TRIG: Qualitative trigonometric ($\mathcal{QT}$) rules. The three angles of a triangle are denoted by $\angle a$, $\angle b$, and $\angle c$, which are opposite to sides A, B, and C, respectively. The left-most column gives the lengths of A and B (in pairs) with respect to a certain reference length of interest, and the top-most row gives the joint angle $\angle c$ between A and B. Each entry of TRIG specifies the corresponding possible length C and $\angle a$ measurements, $< C, \angle a >$. *'l'*= Less, *'sl'*= SlightlyLess, *'eq'*= Equal, *'sg'*= SlightlyGreater, *'g'*= Greater, *'a'*= Acute, *'sa'*= SlightlyAcute, *'r'*= RightAngle, *'so'*= SlightlyObtuse, and *'o'*= Obtuse.

TRIG	a	sa	r	so	o
$<l, l>$	$<l, so{\sim}o>$	$<l{\sim}sl, sa{\sim}o>$	$<l{\sim}sl,sa{\sim}o>$	$<l{\sim}sg,a{\sim}o>$	$<l{\sim}sg,a{\sim}o>$
$<l, sl>$	$<l{\sim}sl, sa{\sim}o>$	$<sl{\sim}sg, a{\sim}so>$	$<sl{\sim}sg,sa{\sim}so>$	$<sl{\sim}sg,sa{\sim}so>$	$<sl{\sim}g,a{\sim}sa>$
$<sl, sl>$	$<l{\sim}sl, sa{\sim}o>$	$<sl{\sim}sg, a{\sim}so>$	$<sl{\sim}sg,a{\sim}sa>$	$<eq{\sim}g,a{\sim}sa>$	$<sg{\sim}g,a>$
$<eq, eq>$	$<l{\sim}eq, sa>$	$<eq{\sim}sg, a>$	$<sg,a>$	$<sg{\sim}g,a>$	$<g,a>$
$<l, eq>$	$<l{\sim}sl, sa>$	$<sl{\sim}sg, a{\sim}sa>$	$<sg,a>$	$<sg,a>$	$<sg{\sim}g,a>$
$<sl, eq>$	$<l{\sim}sl, sa>$	$<eg{\sim}sg, a{\sim}sa>$	$<sg,a>$	$<sg{\sim}g,a>$	$<g,a>$
$<l, sg>$	$<l{\sim}sg, a{\sim}sa>$	$<sl{\sim}g, a{\sim}sa>$	$<sg{\sim}g,a>$	$<sg{\sim}g,a>$	$<sg{\sim}g,a>$
$<sl, sg>$	$<l{\sim}sg, a{\sim}sa>$	$<eq{\sim}g, a{\sim}sa>$	$<sg{\sim}g,a>$	$<sg{\sim}g,a>$	$<g,a>$
$<eq, sg>$	$<l{\sim}sg, a{\sim}sa>$	$<sg, a>$	$<sg{\sim}g,a>$	$<sg{\sim}g,a>$	$<g,a>$
$<sg, sg>$	$<l{\sim}sg, a>$	$<sg{\sim}g, a>$	$<sg{\sim}g,a>$	$<sg{\sim}g,a>$	$<g,a>$
$<sg, g>$	$<l{\sim}g, a>$	$<g, a>$	$<g,a>$	$<g,a>$	$<g,a>$
$<eq, g>$	$<l{\sim}g, a>$	$<sg{\sim}g, a>$	$<g,a>$	$<g,a>$	$<g,a>$
$<sl, g>$	$<l{\sim}g, a>$	$<sg{\sim}g, a>$	$<g,a>$	$<g,a>$	$<g,a>$
$<l, g>$	$<sl{\sim}g, a>$	$<sg{\sim}g, a>$	$<g,a>$	$<g,a>$	$<g,a>$
$<g, g>$	$<l{\sim}g, a>$	$<g, a>$	$<g,a>$	$<g,a>$	$<g,a>$

where $\wedge$ and $\vee$ denote logical operators AND and OR, respectively. $\Longrightarrow$ denotes a logical implication.

The triangular spatial inference rules provide complete descriptions of various types of triangles with respect to the angle space and the qualitative relations as defined in Sections 4.1 and 4.2; hence, they constitute a formal yet qualitative representation of *trigonometry.* These rules can be derived in the following manner:

1. For each of the qualitative primitive values of $< A, B >$ and $\angle c$, identify its numerical bounds.
2. For each combination of the identified bounds, compute and sort the exact values of $< C, \angle a >$.
3. Use the labels of Sections 4.1 and 4.2 to represent the computed values of $< C, \angle a >$.

The above labeling procedure guarantees that the qualitative label representation of the trigonometric relations is *complete* (see Definition 12). It should be noted that the completeness of the qualitative trigonometric rules, as presented in Table 5.1, is only valid with respect to the label semantics developed in Section 4.1 and 4.2. In other words, with a different set of labels and label definitions, some of the rules in the table may have to be rewritten (based again on the above procedure), in order to ensure the completeness of the rules.

Table 5.2. `ADD`: Qualitative addition. Each entry of `ADD` is arranged in a pair of $< length,\ angle >$. The left-most column gives a set of values for $< L_1, \theta_1 >$, and the top-most row indicates a set of values for $< L_2, \theta_2 >$. The remaining entries describe the results of their qualitative addition. Note that label *'nil'* indicates an angle is beyond the $[0, \pi]$ domain, which, based on the qualitative angle definition, is not possible. *'l'*= `Less`, *'sl'*= `SlightlyLess`, *'eq'*= `Equal`, *'sg'*= `SlightlyGreater`, *'g'*= `Greater`, *'a'*= `Acute`, *'sa'*= `SlightlyAcute`, *'r'*= `RightAngle`, *'so'*= `SlightlyObtuse`, and *'o'*= `Obtuse`.

		$< L_2, \theta_2 >$				
	ADD	$<l, a>$	$<sl, sa>$	$<eq, r>$	$<sg, so>$	$<g, o>$
$< L_1, \theta_1 >$	$<l, a>$	$<sg, a{\sim}so>$	$<sg{\sim}g, sa{\sim}o>$	$<g, so{\sim}o>$	$<g, so{\sim}o>$	$<g, o>$
	$<sl, sa>$	$<sg{\sim}g, sa{\sim}o>$	$<sg{\sim}g, o>$	$<g, o>$	$<g, o>$	$<g, nil>$
	$<eq, r>$	$<g, so{\sim}o>$	$<g, o>$	$<g, o>$	$<g, nil>$	$<g, nil>$
	$<sg, so>$	$<g, so{\sim}o>$	$<g, o>$	$<g, nil>$	$<g, nil>$	$<g, nil>$
	$<g, o>$	$<g, o>$	$<g, nil>$	$<g, nil>$	$<g, nil>$	$<g, nil>$

Axiom 5.1.1 (Derivability) *Using the qualitative trigonometric rules of Table 5.1, if among the three sides and three angles in a triangle any three qualitative values are given, the rest can be inferred.*

If each of the qualitative values is interpreted as a *relation*, then each inference rule can also be regarded as a composition rule for new relations.

5.1.2 Qualitative Arithmetic ($\mathcal{QA}$) and Propagation

In spatial analysis, it is often the case that the derivation of the final conclusions may involve combining or subdividing length(s) or angle(s). Such situations occur when two or more triangles are involved and the spatial relationships have to be propagated from one triangle to another. Table 5.2 shows the *qualitative arithmetic* ($\mathcal{QA}$) of both lengths and angles, developed on the basis of the two qualitative quantity space semantics. Each entry of the table is arranged in a pair of $< length,\ angle >$. The left-most column gives one set of values for $<L_1, \theta_1>$, whereas the top-most row indicates another set of values for $<L_2, \theta_2>$. The remaining entries provide the results of qualitative addition, i.e.,

$$< L_3, \theta_3 > \quad = \quad < L_1, \theta_1 > + < L_2, \theta_2 > \tag{5.1}$$

If the qualitative values are viewed as *relations*, the qualitative arithmetic for Euclidean quantities is in fact the *propagation* between the quantities.

5.1.3 Inferencing

Inferencing based on Table 5.1 may not always yield *unique* conclusions. As it is evident from Table 5.1, the combination of qualitative values in the $< C, \angle a >$ entry provides the possible triangular configurations. Each element of $< C, \angle a >$ may contain a sequence of adjacent intervals (primitive values) which indicates the possible quantity domain of a specific variable. Here, it is required that *at least three* qualitative variables (out of six) are given in order to perform the triangular inferences. It should be noted that the three variables can be in any possible combination.

Theorem 4 (Completeness of qualitative trigonometry). *For any triangle, if the values of three of its six qualitative variables are given, then the inferences of the remaining qualitative variables, based on the qualitative trigonometry table, are complete.*

Proof: Since the construction of Table 5.1 is based on the numerical computation of the corresponding variables followed by a qualitative mapping as stated in Definition 9 (i.e., qualitative values are the subsets of the union of mutually disjoint domains or intervals), for each qualitative variable, logical statement 4.14 must be satisfied. Hence, the theorem is proven. ∎

Similarly, the following theorem can be readily proven:

Theorem 5 (Completeness of qualitative trigonometry and arithmetic). *The inferences based on Tables 5.1 and 5.2 are complete.*

Theorem 6 (Minimum requirements for inferencing). *Let n be the number of edges of a non-intersecting polygon. If the polygon is partitioned into a finite set of mutually disjoint sub-polygons, and if at least $2n-3$ qualitative lengths and angles either on or within the boundary of the polygon are qualitatively specified, then all the relative spatial relationships of the polygon can be qualitatively specified.*

Proof: This theorem can be proven by induction.

Basis of induction. Let $n = 3$, where n is the total number of edges. In this case, the corresponding polygon is a triangle. By applying the triangular spatial relationships given in Table 5.1, the remaining unknown spatial relationships can always be inferred from the values of at least $2 \times 3 - 3 = 3$ distinct qualitative variables.

Inductive step. Let $n = k$, where n is the total number of edges in the polygon, and suppose the specification of such a polygon requires at least $2k - 3$ qualitative variables be given. Now, consider a polygon with $k+1$ edges. In this case, without loss of generality, assume that e_i and e_{i+1} are two consecutive edges of the polygon where their interior joint angle is less than π. As shown in Figure 5.1, a diagonal edge is added that connects the two non-common endpoints of e_i and e_{i+1}, as denoted by e'. By doing so, this polygon is subdivided into one triangle $\{e_i, e_{i+1}, e'\}$ and one polygon with k edges (formed by e' and all the original edges except e_i and e_{i+1}). It is assumed that a complete qualitative spatial description of the polygon with k edges (including e') requires at least $2k - 3$ spatial relationships be given. It is also known from Axiom 5.1.1 that when e' is given, a complete description of the triangle requires at least 2 additional spatial relationships. Therefore, at least $2(k+1) - 3$ relationships are required to specify the polygon with $k+1$ edges. Hence, the theorem is proven. ∎

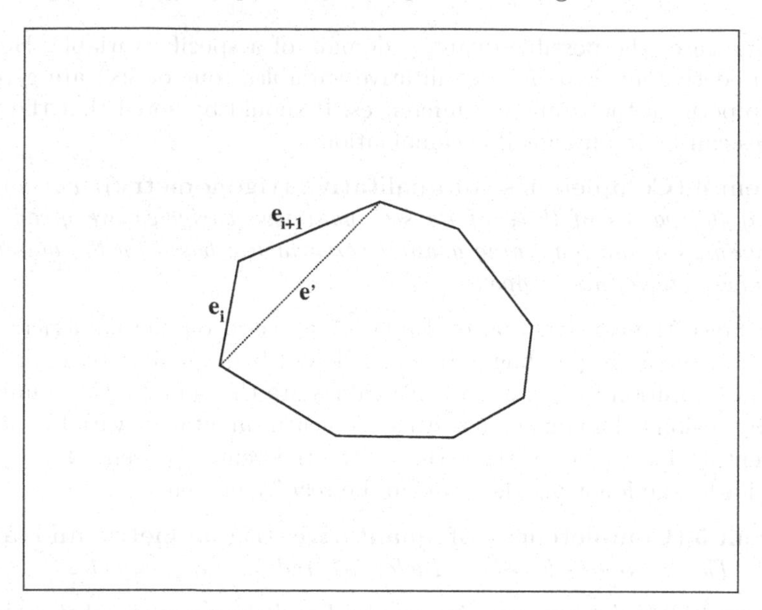

Fig. 5.1. A polygon with $k + 1$ edges can be partitioned into one triangle and one polygon with k edges — the proof of Theorem 6.

Theorem 7 (Completeness of inferencing). *If a non-intersecting polygon has n edges, and $2n - 3$ of its relative spatial relationships are given, then the qualitative spatial inferencing of the rest of its relationships is complete.*

Proof: This theorem can be proven by induction with the same steps as used in the proof of Theorem 6.

Basis of the induction. Let $n = 3$. In this case, the triangular inference rules guarantee that the solution be *complete* (by Theorem 4).

Inductive step. Let $n = k$, and assume that the description of the polygon with k edges (including e') is *complete.* Using the construction, as shown in Figure 5.1, a polygon with $k+1$ edges can be subdivided into one triangle and one polygon with k edges. It is known from the condition that the description of e' must be *complete*, i.e., all the possible values of e' can be inferred. It is also known from Theorem 4 that, with each value of e', and two additional spatial relationships in the triangle, a *complete* description of triangular relationships can be derived. Therefore, the polygon with $k + 1$ edges can be *completely* described. In other words, the *qualitative spatial inferencing* of the spatial relationships in a polygon with $k + 1$ edges is *complete.* ∎

Theorem 8 (Connectedness of qualitative values). [1] *For any continuous variable $x \in \mathcal{R}$ (assume that $\mathcal{R}$ can be subdivided into more than one*

[1] In Appendix A, a grid-based graphical representation of the qualitative trigonometry and qualitative arithmetic rules is presented, where each dimension of a rectangular grid corresponds to a qualitative variable and each cell corresponds to a

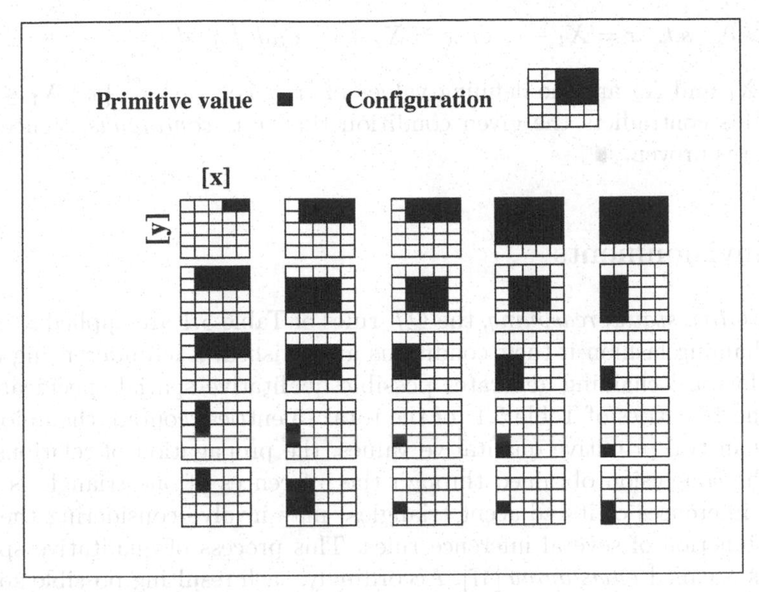

Fig. 5.2. An example of the complete qualitative mapping. The two-dimensional array represents a qualitative mapping of a continuous two-dimensional variable. It can be noted that the cells (corresponding to primitive values) in any two adjacent entries (corresponding to two qualitative values) are connected.

qualitative primitive value), the complete qualitative mapping $[x]$ of x is connected (as illustrated in Figure 5.2), that is:

$$\forall [x_i] \in [X], \exists [x_j] \in [X], s.t. \ q_s \in [x_i] \ \textit{is adjacent to} \ q_t \in [x_j]; \\ q_s, q_t \in \mathcal{Q} - space_{[x]} \tag{5.2}$$

where $[x_i]$ and $[x_j]$ ($[x_i] \neq [x_j]$) are two possible values of x, defined as the subsets of the union of all disjoint partitioning (or primitive) qualitative values q.

For example, if $[x_1]$ = `Acute` $\vee$ `SlightlyAcute`, where $[x_1]$ is composed of two primitive q values: `Acute` and `SlightlyAcute`, then one of the possible values of x connected to $[x_1]$ will be $[x_2]$ = `SlightlyAcute`.

Proof: Since $[x]$ is *complete*, it must map all x values. Assume that there exists a $[x_k]$ which does not satisfy the statement of Theorem 8. From this assumption as well as the definition of qualitative mapping (Definition 9), it is known that $[x_k]$ must be locally bounded, that is,

---

specific value of that variable. From the qualitative grid representation of the $\mathcal{QT}$ rules, as shown in Figure A.1, the following properties can be readily observed: (1) the cells (i.e., qualitative values) within each entry are connected and (2) the cells in one entry are either overlapping or adjacent to the cells in its neighboring entries. In fact, these observations can be proven to be true for any *complete* qualitative representation of continuous variables, as stated in Theorem 8.

$$\exists \epsilon > 0, \quad s.t. \; x = X_1 - \epsilon, \; or \; x = X_2 + \epsilon \quad undefined \tag{5.3}$$

where X_1 and X_2 are the defining values of $[x_k]$, i.e., $[x_k] \stackrel{\text{def}}{=} \{x \mid X_1 \leq x \leq X_2\}$. This contradicts the given condition that x is *continuous*. Hence, the theorem is proven. ∎

5.2 Envisionments

In *qualitative spatial reasoning*, the $\mathcal{QT}$ rules of Table 5.1 are applied in a forward chaining fashion if their conditions are satisfied. Each inferencing cycle of the forward chaining generates possible qualitative spatial specifications. Since the 26 entries of Table 5.1 (of the total 75 entries) contain the unions of more than two primitive qualitative values, the propagation of relationships (e.g., the conclusion obtained through the inferences of one triangle is used for the inferences of its adjacent triangles) may involve considering the logical disjunction of several inference rules. This process of qualitative spatial analysis is called *envisioning* [37]. Accordingly, each resulting possible spatial arrangement is called an *envisionment.*

Envisionments may also be built to serve as a qualitative simulation of configuration change. For example, consider the inference rules given in the first row entries of Table 5.1. The question may be that, given A and B, if $\angle c$ changes from `SlightlyAcute` to `Obtuse`, how does $< C, \angle a >$ change accordingly? Since each entry is a union of some primitive qualitative values, the complete solution requires a process of envisioning as defined in the above paragraph. Each possible envisionment gives a sequence of spatial relationship change defined in terms of primitive qualitative values. From Theorem 8, it follows that these qualitative values must be connected.

5.3 Spatial Planning in $\mathcal{Q}$-Space

The *spatial envisionment* mentioned in the previous section is concerned with finding all the possible sequences of *connected primitive values* in which the next primitive value is either in the same qualitative configuration (a union of primitive values) as the current primitive value or in the next configuration, as shown in Figure 5.2. This is equivalent to saying:

$$\begin{aligned}\{< C_s, \angle a_t > \mid < C_i, \angle a_j > \in [x_n] \; and \; < C_s, \angle a_t > \in [x_n] \\ or \; [x_{n+1}], s = i \pm 1, t = j \pm 1\}\end{aligned} \tag{5.4}$$

where $[x_n]$ and $[x_{n+1}]$ denote the current and the next qualitative configurations, respectively. This is essentially a problem of searching for *all possible routes* connecting an *initial* primitive value to a *goal* value in the *given qualitative quantity space.* There may be more than one initial or goal value. Here, the cost of transition from one primitive value to its adjacent values is not considered.

In this section, the above problem is revised to take into account the cost of a possible route:

Given an initial and a goal configurations and a valid domain of $\mathcal{Q}$-space, the objective is to find a sequence of configuration transitions that connects the given initial and goal configurations. In particular, one instance of this problem is considered — qualitatively optimal location connection *in free-space $\mathcal{E} - \mathcal{P}$.*

5.3.1 Qualitative Route

Given a graphic representation of qualitative locations in $\mathcal{E} - \mathcal{P}$, a *qualitative route* from node N_0 to node N_k can be formally defined as a sequence $\{N_0, A_0, N_1, A_1, ..., A_{k-1}, N_k\}$ where, for each i, the endpoints of arc A_i are N_i and N_{i+1}. The *length* of the route is defined by the number of arcs it contains, which corresponds to the number of m-closure regions traveled; or

$$[length_i] = [length_{(i-1)}] + 1 \tag{5.5}$$

5.3.2 Clearance Measurements of a Qualitative Route

This subsection formally defines the clearance as well as orientation cost associated with a qualitative route.

Suppose that a convex polygonal object is to be moved from one location to another in a polygonal environment represented as a set of disjoint m-closure regions. By definition, the length of an m-edge measures the maximum width provided for the polygon to pass through by translation. Thus, for an m-closure region, the relative clearance of translation through each of its m-edges can be computed. As used in the route search, this measurement promotes wide paths for an object to translate and sweep. Based on such an observation, the notion of *passage clearance* can be defined.

Definition 16 (Passage clearance). *The passage clearance of a qualitative route at an m-edge is defined as:*

$$[Clearance(i)] = [d_{\mathrm{m-edge_i}} \mid d_{\mathrm{ref}}] \tag{5.6}$$

where $[d_{\mathrm{m-edge_i}} \mid d_{\mathrm{ref}}]$ denotes the qualitative m-edge distance with respect to a reference distance, d_{ref}.

From the definition of m-closure region (see Definition 14), it is known that if each of the individual polygons within an environment shrinks into a point, then the set of m-closure regions which partitions the environment will accordingly degenerate into a set of convex polygons. Thus, when a polygonal environment is *less cluttered*, the properties of a convex polygon, especially the property of convex chain existence, may be analogously applied to the m-closure regions of the environment in order to establish the notion of *orientation cost* of a qualitative route.

Theorem 9 (Existence of a convex chain). *For any interior point of a convex polygon, it is always possible to construct an open-loop convex chain, from one edge to another, which passes through that interior point.*

Proof: Assume that e_j and e_k are the two edges through which a convex chain is to be constructed, and q is the interior point that the convex chain has to pass through. Construct two rays from q which pass through e_j and e_k. Based on the convexity of the polygon, it is known that each ray must intersect the boundary of the polygon only once. Let the intersection points with e_j and e_k be z_j and z_k, respectively. Hence, a convex chain $\{z_j, q, z_k\}$ is constructed. ∎

Theorem 10 (Angular ordering of a convex polygon). *The vertices in a convex polygon occur in an angular order about any interior point.*

Proof: Let P be a convex polygon and q be an arbitrary point lying inside the polygon. By the definition of the convex polygon, it is known that if a ray is drawn from q toward the boundary of the polygon, there must exist only one intersecting point. Assume that p_{i-1}, p_i, and p_{i+1} are three consecutive vertices on P which are *not* in an angular order. By constructing the rays from the interior point q through these three vertices, as shown in Figure 5.3, it can be found that one of the three rays will intersect with the boundary of the polygon twice. This contradicts the convexity of P. ∎

Theorem 11 (Sweeping angles in a convex chain). *If $\{q_1, q_2, ..., q_{i-1}, q_i, q_{i+1}, ..., q_n\}$ is a convex chain, and the angle change from edge $\overline{q_i q_{i+1}}$ to edge $\overline{q_{i+1} q_{i+2}}$ is $\triangle\theta_{i,i+1} > 0$, then $\triangle\theta_{1,n-1} > \triangle\theta_{i,i+1}$, where $i = 1, 2, ..., n-2$.*

Proof: Starting from q_1, traverse the convex chain in a counter-clockwise direction, as shown in Figure 5.4. Without loss of generality, assume that the counter-clockwise sweeping angle is *positive*. It can be implied from Theorem 10 that, as the convex chain is matched around, each angle change from the current direction to the next direction must satisfy:

$$0 < \triangle\theta_{i,i+1} < \pi \tag{5.7}$$

where $\triangle\theta_{i,i+1}$ denotes the angle change when turning from edge $\overline{q_i q_{i+1}}$ to edge $\overline{q_{i+1} q_{i+2}}$.

If all the angle changes are summed up, then

$$\sum \triangle\theta = \triangle\theta_{1,n-1} > \triangle\theta_{i,i+1} \tag{5.8}$$

where $i = 1, 2, ..., n-2$. Hence, the theorem is proven. ∎

Based on the assumption that the polygonal environment is less cluttered, it is known from Theorems 9 and 11 that a convex chain can be constructed

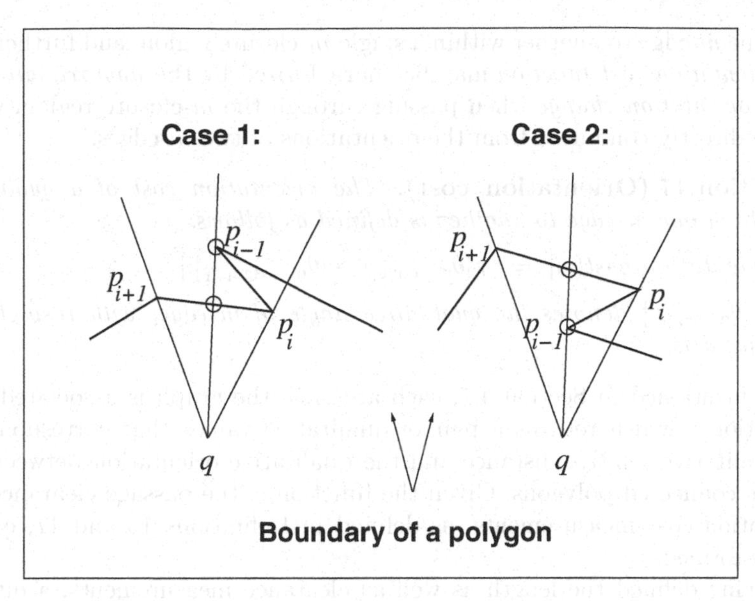

Fig. 5.3. Let p_{i-1}, p_i, and p_{i+1} be three consecutive vertices on the boundary of a polygon. If p_{i-1}, p_i, and p_{i+1} are not in an angular order, one of the three rays from q must intersect with the boundary of the polygon twice — the proof of Theorem 10.

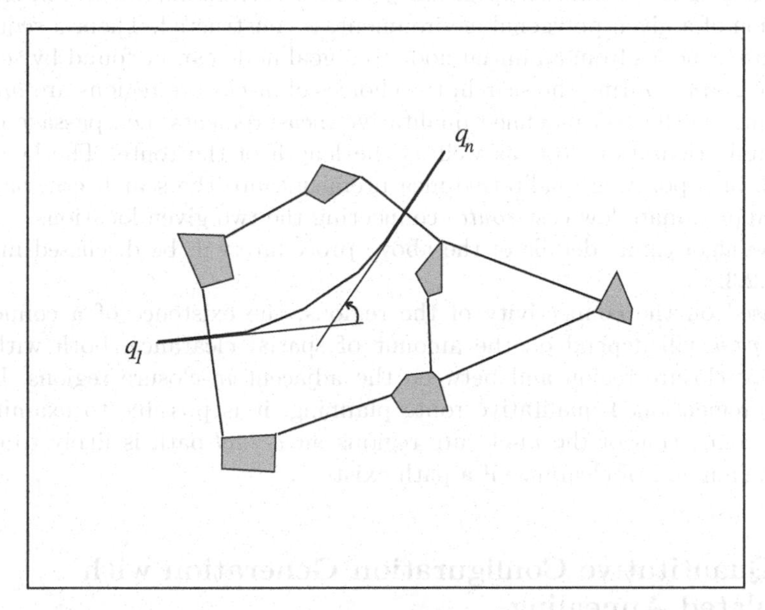

Fig. 5.4. The total angle changed after the traversal from q_1 to q_n is greater than the angle change of the subchain — the proof of Theorem 11.

from one m-edge to another within a single m-closure region, and furthermore the *orientation cost function* may be characterized by the *approximate maximum orientation change* when passing through the m-closure region, which can be directly computed from the orientations of two m-edges.

Definition 17 (Orientation cost). *The orientation cost of a qualitative route from one m-edge to another is defined as follows:*

$$[Orientation\_cost(i)] = | [\theta_{\mathrm{m-edge_i}}] - [\theta_{\mathrm{m-edge_{i-1}}}] | \quad (5.9)$$

where $[\theta_{\mathrm{m-edge_i}}]$ *denotes the qualitative angle of* m-*edge*$_i$ *with respect to a reference axis.*

As mentioned in Section 4.7, each arc A in the graph is associated with a function c which returns a pair of qualitative values that corresponds to the qualitative relative distance and the qualitative orientation between two m-edge connected polygons. Given the function c, the passage clearance and orientation cost measurements, as defined in Definitions 16 and 17, can be readily formed.

Having defined the length as well as clearance measurements, a qualitative route connecting two locations with the optimalities of minimum length, maximum clearance, and minimum orientation cost can be planned as follows: First, a weighted connectivity graph $\mathcal{G}$, which corresponds to the m-closure partition of a given polygonal environment, is constructed. Then, a sequence of adjacent nodes from an initial node to a goal node can be found by searching the graph. During the search, the choices of m-closure regions are *ordered* according to the above-defined qualitative measurements, i.e., passage clearance and orientation cost, as well as the length of the route. The heuristic search, incorporating qualitative measurements into the search criteria, will yield approximate low-cost *routes* connecting the two given locations.

The algorithmic details of the above procedure will be discussed in Section 8.2.3.

Based on the connectivity of the regions, the existence of a connected *exact path* will depend on the amount of spatial clearance, both within a single m-closure region and between the adjacent m-closure regions. Using the aforementioned qualitative route planning, it is possible to examine in which connection of the m-closure regions an *actual* path is likely to exist, and the amount of clearance if a path exists.

5.4 Quantitative Configuration Generation with Simulated Annealing

Suppose that $\mathbf{C}_{\mathrm{init}}$ and $\mathbf{C}_{\mathrm{goal}}$ are initial and goal configurations, respectively, and a path between them is to be found. The most obvious way of using the Euclidean distance as a criterion (an objective function, E) in an optimization search is to begin with the initial configuration and search its neighboring configurations one by one. Changes that improve (decrease) the distance function are accepted and others are rejected. This iterative strategy is one of the standard search techniques and suffers from some difficulties. For instance, by requiring that the iterative steps move only "downhill" on the (usually multidimensional) objective function (E), it is almost certain that the search will get trapped in a local minimum.

As illustrated by Glavina [72], in A^* search with a Euclidean distance as its heuristic function, a local-minimum configuration is inevitable and can be escaped only by backtracking throughout neighboring configurations. Such a phenomenon may be described as *dead-end bag filling*, as shown in Figure 5.5.

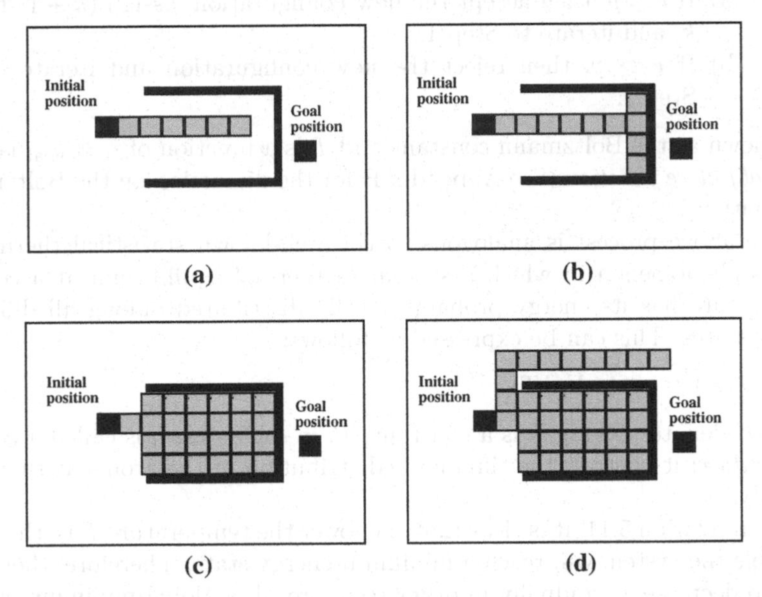

Fig. 5.5. The standard informed search for an *exact* path, such as A^* search with a Euclidean distance as its heuristic function, fills up dead-ends (or local minima) (based on the illustration of Glavina [72]).

What follows describes a well-known randomized search technique to solve the above problem of getting trapped in a local minimum and to trade exactness for efficiency. The technique is analogous to a statistical thermodynamic process and is hence called *simulated annealing.*

The key to simulated annealing is a Monte Carlo algorithm as developed by Metropolis et al. [152]. This algorithm generates a sequence of states (or configurations). The sequence is a Markov chain in which each state depends on the previous one according to the Boltzmann distribution. By using Metropolis's algorithm, the above stated distance-minimization problem can be solved in an iterative fashion as follows:

Algorithm ANNEAL

1. Make some random change to configuration $\mathbf{C}_\mathrm{i}$, resulting in $\mathbf{C}_{(\mathrm{i+1})}$ and compute a distance, $E_{(\mathrm{i+1}),\mathrm{goal}}$, between the new configuration $\mathbf{C}_{(\mathrm{i+1})}$ and the goal configuration $\mathbf{C}_{\mathrm{goal}}$.
2. If $E_{(\mathrm{i+1}),\mathrm{goal}} \leq E_{\mathrm{i},\mathrm{goal}}$, then accept $\mathbf{C}_{(\mathrm{i+1})}$ and assign $(i+1)$ to i; iterate the previous step.
3. If $E_{(\mathrm{i+1}),\mathrm{goal}} > E_{\mathrm{i},\mathrm{goal}}$, then compute the following probability:
$$p = \exp(-(E_{(\mathrm{i+1}),\mathrm{goal}} - E_{\mathrm{i},\mathrm{goal}})/kT) \tag{5.10}$$
and generate a random number $r \in [0, 1]$.
   a) If $r \leq p$, then accept the new configuration, assign $(k+1)$ to k, and iterate to Step 1.
   b) If $r > p$, then reject the new configuration and iterate to Step 1.

k is known as the Boltzmann constant and T is a function of i. $E_{\mathrm{i},\mathrm{goal}}$ is used as an *objective function* (See Appendix B for the discussion on the Boltzmann distribution).

The above process is analogous to the well-known statistical thermodynamics phenomenon in which a system in thermal equilibrium at a certain temperature has its energy probabilistically distributed among all different energy states. This can be expressed as follows:

$$Prob(E) \propto \exp(-E/kT) \tag{5.11}$$

where E denotes *energy*, k is a constant. T, in such a case, is called a *control temperature*; it controls the time for redistribution of the atoms as they lose mobility.

From relation 5.11, it is clear that the lower the temperature T is, the more probable the system will reach a minimum-energy state. Therefore, there is a need to decrease T gradually in order to approach a global minimum-energy state. This process is also known as the *scheduled cooling* process which leads to a crystalline and ordered state of global minimum energy [110].

Based on the above theory, in the rest of the book the parameter T in the ANNEAL algorithm will be referred to as a control temperature, and the objective evaluation function E will sometimes be referred to as an energy function.

6

How to Reason about Mechanism Configurations

Traditionally, a linkage kinematic problem has been solved using either analytical (including constraint) methods, graphical methods, or numerical simulation methods [80, 87, 90, 183]. These methods are mathematically complete, but require detailed formulation of the problem, which is often linkage-dependent [91, 170]. For the problem of deriving the *locus* traced by a certain point of the mechanism, the standard algebraic method has been extensively used in the past (a detailed review is provided in [91]). In the algebraic formulation, this problem is translated into that of finding a point (x, y) which traces the curve satisfying an equation $F(x, y) = 0$, where F is a rational integral algebraic function of x and y.

All these methods rely on precise mathematical analysis, which in turn requires the knowledge of *exact mechanism geometry.* In real-life situations, however, the precise geometry of a mechanism being studied may not be given as *a priori* knowledge. Taking robot *task* planning as an example, the available geometric information about a mechanism to be manipulated is in some cases only approximate in nature, and not accurate.

This chapter introduces a method of deriving the loci of points in a one-degree-of-freedom linkage mechanism, which applies the qualitative spatial inferencing and randomized search techniques as described in Sections 5.1 and 5.4, respectively. In this method, the input mechanism geometry may be given only in qualitative terms.

6.1 An Overview of the Method

In the proposed method, the locus of constrained mechanisms is analyzed in two stages. The first stage is to perform the qualitative spatial inferencing with trigonometric rules (i.e., $\mathcal{QT}$ rules), as presented in Section 5.1.1, and derive the qualitative spatial description of the position of any given link. The second stage is to apply the randomized local search technique given the qualitative position descriptions, and generate the quantitative configuration of the mechanism. Thus, instead of analytically computing the next configuration or graphically representing the velocities as in classical kinematics, this method is mainly based on qualitative spatial analysis and simulated annealing. As will be discussed later in this chapter, both stages are linkage-independent.

The overall method is outlined in algorithm C_INSTANT_CONFIG. To study a complex linkage, this algorithm decomposes the linkage into a set of *independent four-bar linkages.*

Definition 18 (Independent four-bar linkages). *A four-bar linkage is considered independent if and only if it has at most one link that is shared with other four-bar linkages.*

Algorithm C_INSTANT_CONFIG

Input: A linkage with its dimensions and one of its joint angles (corresponding to the input driver link).
Output: A sequence of instantaneous configurations of the linkage corresponding to the angular displacement of the driver link.

1. **Linkage decomposition:** Represent a given linkage as the composition of a set of *independent* sub-linkages (each one equivalent to a four-link mechanism). If this is not possible, then exit; else, find the sub-linkage which contains the input driver link.
2. **Qualitative labeling:** Derive the qualitative lengths of the links with respect to the assumed fixed link of the linkage and the current qualitative angular position of the driver link. These qualitative spatial relationships are determined according to the semantics given in Sections 4.1 and 4.2.
3. **Envisioning:** Apply algorithm QUALITATIVE_CONFIG in each sub-linkage (i.e., qualitative spatial inferencing) to decide within which qualitative region the quantitative configuration lies (i.e., to derive a qualitative configuration).
4. **Local search:** Apply algorithm ANNEAL (see Section 5.4) to yield a quantitative configuration within the predicted qualitative region.
5. If an associated sub-linkage is not traversed, then go to Step 3 (**Envisioning**).
6. Repeat Steps 3 (**Envisioning**) to 5 for all the angular positions of the driver link.

The following sections describe in details Steps 3 (**Envisioning**) and 4 (**Local search**) with illustrative examples.

6.2 Qualitative Configuration Analysis

An instantaneous *configuration* of a mechanism is defined by the instantaneous positions of the individual links in the mechanism. In kinematic analysis, if a linkage mechanism is represented as a collection of connected line segments, then its configurations can simply be described in terms of the lengths of line segments, L_i, and their joint angles, θ. Thus, by triangulating the connected line segments and using qualitative trigonometry, it is possible to describe the qualitative configuration of the mechanism (see Theorem 6).

What follows is an outline of the general algorithm for performing qualitative configuration analysis of *four-bar linkage* mechanisms (or equivalent) that solves the following problem:

> *Given the qualitative lengths of links as well as the qualitative angular position of a driver link, find the positions of other links in terms of their joint angles.*

Let the qualitative dimensions of a specific linkage be given (θ_{ij} denotes the joint angle between links L_i and L_j). The following algorithm generates as an output the qualitative configuration of the linkage expressed in terms of joint angles (see Figure 6.1).

Algorithm `QUALITATIVE_CONFIG`

1. **Inferencing:** If the sub-linkage contains a slider, then directly infer the configuration with $\mathcal{QT}$ rules, and exit; else form an intermediate link L_a between the non-adjacent end-points of the driver link L_2 and its reference link L_1, and for each of the two triangles perform the same steps as in the previous case.
2. Derive θ_{34} based on the ordering of L_a, L_3, and L_4, and similarly infer θ_{3a} and θ_{4a}.
3. Determine θ_{14} and θ_{23} with $\mathcal{QA}$ rules. If it is a non-crossing configuration, then $\theta_{14} = \theta_{1a} + \theta_{4a}$ and $\theta_{23} = \theta_{2a} + \theta_{3a}$ (see Figures 6.1a and 6.1b). Otherwise, $\theta_{14} = |\theta_{1a} - \theta_{4a}|$ and $\theta_{23} = |\theta_{2a} - \theta_{3a}|$ (see Figures 6.1c and 6.1d).

Note that "[ ]" signs around the qualitative variables have been omitted in the presentation.

6.2.1 Examples

The method of *qualitative* configuration analysis can be best illustrated in the following examples using the `QUALITATIVE_CONFIG` algorithm[1] (see Figure 6.2).

[1] The qualitative configuration analysis of linkages is carried out by a program developed using the NASA/CLIPS rule-based prototyping tool.

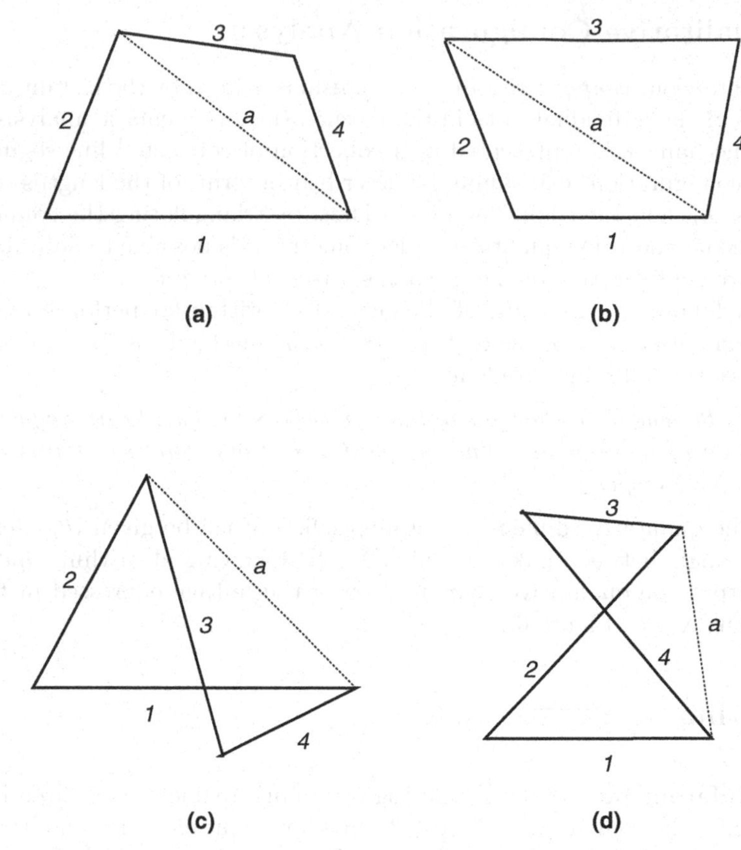

Fig. 6.1. Closed-chain four-bar linkages with different dimensions or configurations, as referred to in algorithm QUALITATIVE_CONFIG. It is assumed that link L_1 is stationary with respect to a fixed reference frame.

Example 6.1: Instantaneous Configuration of A Four-Bar Linkage

In the four-bar linkage shown in Figure 6.2, the partial ordering of the lengths of individual links in the mechanism as well as the qualitative value of an input angle, θ_{21}, between L_2 and L_1 are given (note that it is assumed that link L_2 is the driver link). These qualitative spatial relationships may be stated as follows:

$$[\theta_{21}] = \texttt{Obtuse} \tag{6.1}$$
$$[L_1 \mid L_1] = \texttt{Equal} \tag{6.2}$$
$$[L_2 \mid L_1] = \texttt{Equal} \tag{6.3}$$
$$[L_3 \mid L_1] = \texttt{Greater} \tag{6.4}$$
$$[L_4 \mid L_1] = \texttt{SlightlyGreater} \tag{6.5}$$

The instantaneous angular position, $[\theta_{41}]$, is desired.

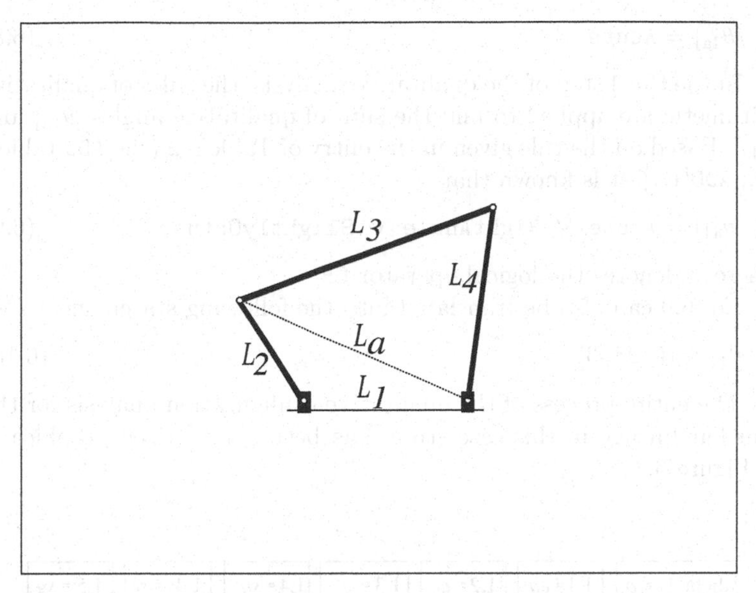

Fig. 6.2. Example 6.1. A closed-chain four-bar linkage as used in the illustration of the QUALITATIVE_CONFIG algorithm. It is assumed that link L_1 is fixed with respect to a reference frame, and link L_2 is the driver link. Link L_a is added for the purpose of qualitative configuration analysis.

Based on algorithm QUALITATIVE_CONFIG, the qualitative analysis directly proceeds with the **Inferencing** step. In doing so, an intermediate link L_a is added between the non-adjacent end-points of L_2 and L_1, as shown in Figure 6.2. Thus, by applying the $\mathcal{QT}$ rule given in Table 5.1 (the TRIG table), i.e., the entry in the 4th row and the 5th column, denoted by TRIG[4,5], the following can be obtained:

$$[\theta_{1a}] = \texttt{Acute} \tag{6.6}$$

$$[L_a \mid L_1] = \texttt{Greater} \tag{6.7}$$

The next inferencing step is concerned with the triangle formed by links L_a, L_3, and L_4, which have the qualitative lengths Greater, Greater, and SlightlyGreater, respectively. The goal is to find the qualitative joint angle, $[\theta_{4a}]$, between links L_4 and L_a. In doing so, the qualitative trigonometric rules of Table 5.1 are applied. Unlike the analysis of $\triangle L_a L_2 L_1$ where two link lengths and one joint angle are known, here the three lengths are given. Without loss of generality, links L_3 and L_a are considered as A and B (as in Table 5.1), respectively, and the entries are found from the row determined by the values of A and B which satisfy the constraints of C (i.e., L_4). As a result, entry TRIG[15,1] is accepted as a qualitative description of the triangle $\triangle L_a L_3 L_4$. This step leads to the determination of $[\theta_{4a}]$, that is,

$$[\theta_{4a}] = \texttt{Acute} \tag{6.8}$$

In the third step of the qualitative analysis, the rules of qualitative arithmetic are applied to find the sum of qualitative angles $[\theta_{1a}]$ and $[\theta_{4a}]$. Based on the rule given in the entry of Table 5.2 (the ADD table), i.e., ADD[1,1], it is known that

$$[\theta_{41}] = \texttt{Acute} \vee \texttt{RightAngle} \vee \texttt{SlightlyObtuse} \tag{6.9}$$

where $\vee$ denotes the logical operator OR.

Eq. 6.9 can also be translated into the following statement:

$$\theta_{41} \in (0, 2\pi/3] \tag{6.10}$$

The entire process of the qualitative configuration analysis for the four-bar linkage in this case study has been summarized graphically in Figure 6.3.

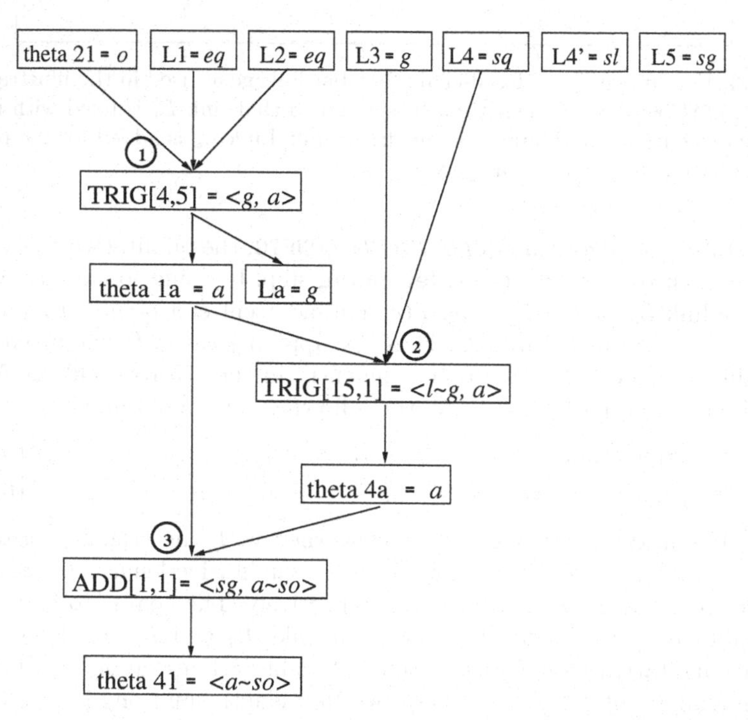

Fig. 6.3. A summary of spatial inferencing steps in deriving the qualitative configuration of the constrained linkage mechanism as given in Figure 6.2 (Example 6.1). The circled *numbers* indicate inferencing steps. TRIG$[i, j]$ and ADD$[i, j]$ denote the inference rules given in the ith row and the jth column of the *qualitative trigonometry* and *qualitative addition* tables, respectively.

Example 6.2: Instantaneous Configuration of a Complex Linkage

Example 6.1 has illustrated how to qualitatively describe the instantaneous configurations of a *single closed-chain four-bar linkage.* Such reasoning plays an essential role in the analysis of complex linkages. As stated in algorithm C_INSTANT_CONFIG, if the step of linkage decomposition succeeds, the problem of analyzing a complex linkage can be reduced to that of analyzing a set of connected four-bar linkages.

This example considers a complex linkage, as shown in Figure 6.4. In this mechanism, links L_1 and L_7 are both fixed with respect to a frame of reference. All the kinematic joints are revolute joints except the joint between links L_6 and L_7. Since link L_6 slides along link L_7, these two links form a sliding joint.

For such a mechanism, the first step (**Linkage decomposition** step) of C_INSTANT_CONFIG proceeds as follows: First, the mechanism is represented as a connectivity graph where each node denotes a joint and each arc denotes a link. From this graph, the independent four-bar linkages of this mechanism are identified.

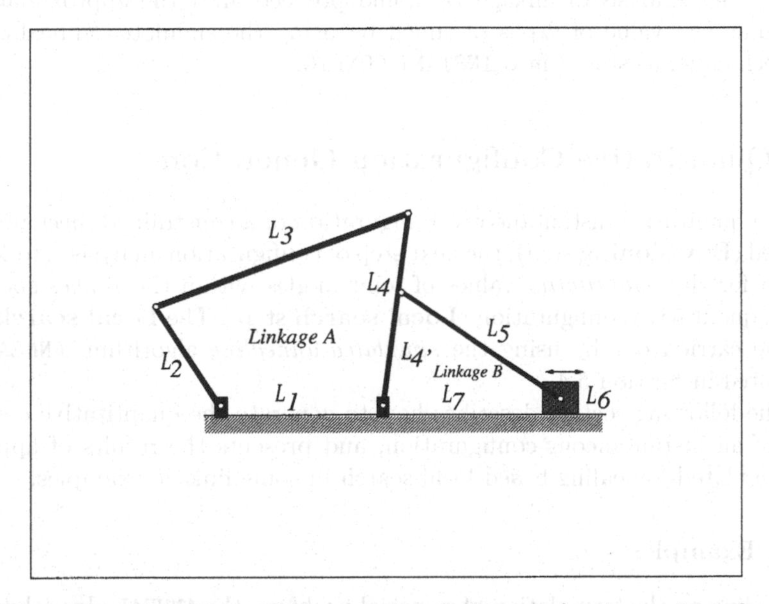

Fig. 6.4. Example 6.2. The analysis of this mechanism using algorithm C_INSTANT_CONFIG proceeds with a step of decomposing the mechanism into a set of independent sub-linkages (equivalent to closed-chain four-bar linkages). In this mechanism, links L_1 and L_7 are both fixed with respect to a frame of reference. All the kinematic joints, except the joint between links L_6 and L_7 (a sliding joint), are revolute joints.

In this example, two four-bar linkages within the mechanism can be identified, one of them being a special case of a four-bar linkage with three rotational degrees and one translational degree of freedom.

The identified four-bar linkage on the left-hand side, called linkage A, is in fact the same as the one studied in the previous section (see Figure 6.2). Assume that all the qualitative geometric information about this four-bar linkage remains the same. In addition, with respect to the four-bar linkage on the right-hand side, called linkage B, the following measurements are given:

$$[L'_4 \mid L_1] = \texttt{SlightlyLess} \qquad (6.11)$$

$$[L_5 \mid L_1] = \texttt{SlightlyGreater} \qquad (6.12)$$

Given the above conditions, the qualitative configuration of linkage B can be easily described using the `QUALITATIVE_CONFIG` algorithm. In doing so, the shared link between the two linkages will be chosen as a new driver link, and hence the steps for linkage A can be repeated. However, it should be noted that in this case the equivalent follower link L_6 becomes a single point (i.e., the linkage forms a triangle), therefore no intermediate link is added.

The analysis of linkage B should proceed after the approximate numerical value of θ_{41} is obtained by using the simulated-annealing technique, as stated in `C_INSTANT_CONFIG`.

6.3 Quantitative Configuration Generation

Once a qualitative instantaneous configuration of a constrained mechanism is derived (**Envisioning** step), the next step of configuration analysis is to *locally* search for the *quantitative* values of joint angles within the *ranges* specified in the qualitative configuration (**Local search** step). The **Local search** step can be carried out by using the *simulated-annealing* algorithm, `ANNEAL`, as presented in Section 5.4.

The following section describes how to generate the quantitative description of an instantaneous configuration, and presents the results of applying the simulated-annealing-based local search in some linkage examples.

6.3.1 Examples

Depending on the formulation of a spatial problem, the `ANNEAL` algorithm may be applied in different ways. In the following examples of *four-bar linkages*, the **quantitative configuration search**[2] problem is formulated as follows:

[2] The `ANNEAL`-based quantitative configuration generation is implemented in MathWorks's MATLAB. The program takes a qualitative configuration description as an input, performs local simulated-annealing search, and plots the accepted quantitative configurations.

From the *qualitative lengths* of a *driver* and a *follower* as well as the *inferred qualitative descriptions* of their *joint angles* (with respect to a fixed link of the linkage), find the corresponding *numerical ranges* of the lengths and joint angles (based on the semantics given in Definitions 10 and 11), and *search* for the *optimal quantitative lengths and joint angles* within those ranges, *such that* the lengths of individual links in the obtained configuration have the same qualitative values as those in the *initial* configuration.

Example 6.3: Quantitative Description of Linkage Paths

This example revisits the four-bar linkage of Figure 6.2, and examines the **Local search** of its instantaneous configurations (as stated in the `C_INSTANT_CONFIG` algorithm). Such a search step is mainly concerned with finding the *quantitative* configuration description based on its qualitative description.

Quantitative Description of an Instantaneous Configuration

Based on the above problem formulation (i.e., the **quantitative configuration search** problem), the following *configuration variable*, $\mathbf{C}_i$, can be identified and used for the simulated-annealing-based local search (see algorithm `ANNEAL` of Section 5.4):

$$\mathbf{C}_i = (\theta_{12}, \theta_{41}, L_2, L_4) \tag{6.13}$$

where θ_{12} and θ_{41} denote the joint angles formed by links L_1 and L_2 and by links L_1 and L_4, respectively. Both joint angles are specified with respect to the fixed link L_1 of the mechanism. It should be noted that the length of link L_1 is used as the reference length of a qualitative length, and hence it is assumed to be a constant. On the other hand, link L_3 as well as joint angles θ_{32} and θ_{34} can be determined if $\mathbf{C}_i$ is known, and thus they are not included in the configuration variable representation.

During each iteration of the `ANNEAL` algorithm, a quantitative configuration, $\mathbf{C}_i$, is selected within the numerical bounds *mapped* from the qualitative spatial descriptions. In other words, the qualitative joint angles derived from the spatial inferencing will be used to place the limits over the *search space* for the actual numerical angles in $\mathbf{C}_i$. For instance, if $[\theta]_{next}$ = `SlightlyAcute`, then the actual θ will be bounded by $[\pi/3, \pi/2)$, based on the definition of the `SlightlyAcute` label.

The objective function to be minimized (i.e., the E function in the `ANNEAL` algorithm) is defined as the sum of the differences between two qualitative lengths, one in the current (generated) configuration and the other in the initial configuration:

$$E = \sum_j \mid [U_j] - [u_j] \mid \; = \; \sum_j \Delta[U_j] \tag{6.14}$$

where $[U_j]$ is the qualitative length of the jth link (L_j), as computed from the *current configuration* $\mathbf{C}_i$. $[u_j]$ is the qualitative length of L_j, as given in the *initial qualitative configuration.*

$\Delta[U_j]$ can be computed in the following way: *If* both $[U_j]$ and $[u_j]$ have the *same qualitative value*, then $\Delta[U_j] \stackrel{\text{def}}{=} 0$; *otherwise*, $\Delta[U_j] \stackrel{\text{def}}{=} \min\{\mid U_j - a \mid, \;\; \mid U_j - b \mid\}$, where a and b are the *upper* and *lower* bounds of the given $[u_j]$, respectively.

After *several iterations* within ANNEAL, an *approximate quantitative configuration* will be *accepted* from the annealing process if the objective function E of the current configuration has reached a certain minimum *threshold.* That is also to say, a configuration $\mathbf{C}_i$ is considered as the quantitative representation of an instantaneous configuration if the values of the qualitative lengths in the obtained configuration are *close enough to* those in the initial configuration (see the definition of the E function in Eq. 6.14). The accepted $\mathbf{C}_i$ is sometimes called an *acceptable quantitative configuration.*

Constrained Path Derivation

The constrained path analysis of a certain point in a mechanism can be performed as follows: First, a sequence of *input angular positions* is identified for the driver link. These positions are specified using the angle labels as defined in Definition 11. Figure 6.5 presents the sequence of angular positions used in this example (ordered in a counter-clockwise direction).

Next, the *qualitative* description of mechanism configuration is inferred using QUALITATIVE_CONFIG (i.e., the **Envisioning** step in the C_INSTANT_CONFIG algorithm), as has already been illustrated in Example 6.1. The local search for an approximate quantitative configuration is then performed using ANNEAL (i.e., the **Local search** step in the C_INSTANT_CONFIG algorithm) as soon as the qualitative description is derived. The above steps are repeated until all the input angular positions have been traversed.

Given the *qualitative lengths* in an initial configuration as given in Example 6.1, and a sequence of driver positions as given in Figure 6.5, a sequence of linkage configurations can be readily generated. Figure 6.6 presents a set of eleven (11) approximate *quantitative* configurations (including the initial configuration), ordered with respect to the driver positions, as shown in Figure 6.5.

If the position of a certain point on the linkage is recorded at each obtained instantaneous configuration, a *constrained path* pertaining to the specific point can be constructed. Figure 6.7 shows the path of a

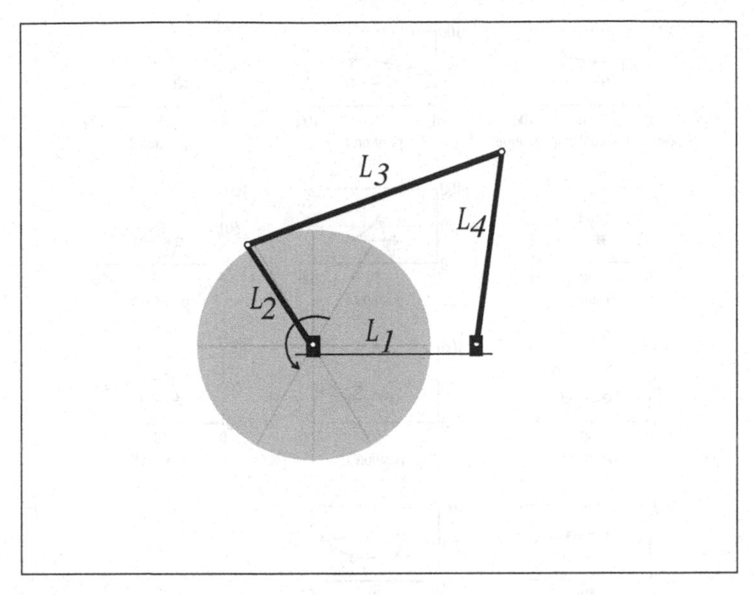

Fig. 6.5. Example 6.3. Given a sequence of *qualitative angular positions* for *driver* link L_2, the corresponding exact instantaneous configurations of the linkage at each step may be derived. It is assumed that link L_1 is fixed with respect to a reference frame.

mid-point on the floating link L_3 (or *coupler*) of the four-bar linkage, where the 11 positions of the mid-point are interpolated using *cubic spline fits.*

A Comparison with the Algebraically Derived Path

The path generated from a set of qualitative spatial descriptions may be compared to the one computed from an *algebraic* procedure *if the precise lengths of the same four-bar linkage are provided.* One of the algebraic methods of computing such a path is to relate the joint angles with the coordinates of a position, as formulated by Hunt [91].

Suppose that the *exact lengths* of the four-bar linkage presented in Figure 6.5 resemble the *actual lengths* of the linkage. It is possible to compute the path of the mid-point on the floating link using Hunt's algebraic method. Figure 6.8 shows the path of such a point constructed by interpolating twenty (20) algebraically computed positions. Comparing the path of Figure 6.7 to the one of Figure 6.8, it can be noted that the two paths, although not exactly equivalent, are similar to each other in terms of the *topology* and *general shape.*

As has been pointed out in Section 5.4, the $T(i)$ *function* of Eq. 5.10 for the annealing process (also known as an *annealing schedule*) will affect

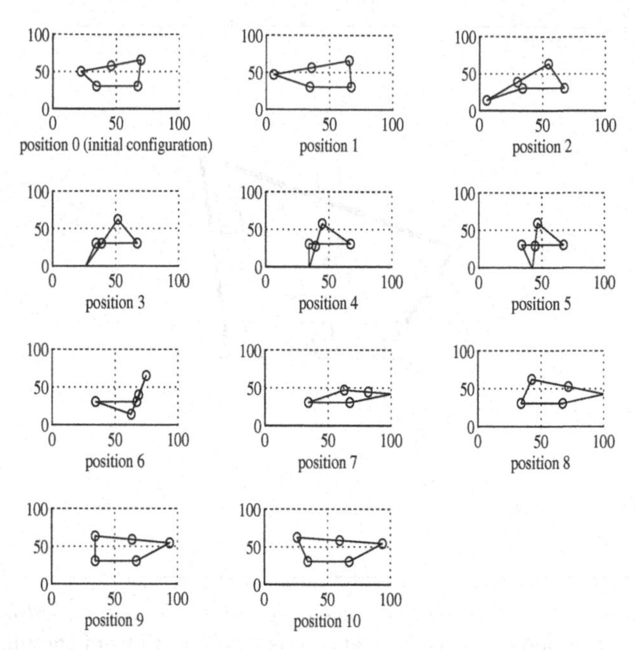

Fig. 6.6. For the linkage mechanism, as shown in Figure 6.2, *a sequence of instantaneous configurations* can be generated when all the lengths are expressed in *qualitative terms.* Each spatial configuration is generated in two steps: (1) qualitative spatial inferencing based on algorithm QUALITATIVE_CONFIG and (2) simulated-annealing-based local search within the ranges of the inferred qualitative spatial variables (e.g., joint angles). In each plot, the bottom horizontal link (called link L_1) is fixed with respect to a reference frame. Link L_2 (numbered clockwise from L_1) is the driver link, which rotates in a counter-clockwise direction, from its initial instantaneous position through a sequence of qualitative angular displacements, as specified in Figure 6.5.

the speed of reaching an *acceptable quantitative configuration.* Some of the frequently applied annealing schedules are as follows (see [9, 16, 79, 146, 154] for detailed discussions):

1. **Constant annealing:** $T(i) = K$;
2. **Arithmetic annealing:** $T(i) = T(i-1) - K$;
3. **Logarithmic annealing:** $T(i) = K/(ln(i+1))$;
4. **Inverse annealing:** $T(i) = K/(i+1)$; and
5. **Exponential annealing:** $T(i) = T(i-1)exp(-T(i-1)/\sigma)$;

where i is the current search step, K is a constant, and σ is the standard deviation of the energy function E (the objective function).

In the experiments of the current work, all the above schedules have been tested, and the **Exponential annealing** schedule has been found to be the most efficient in approaching an equilibrium.

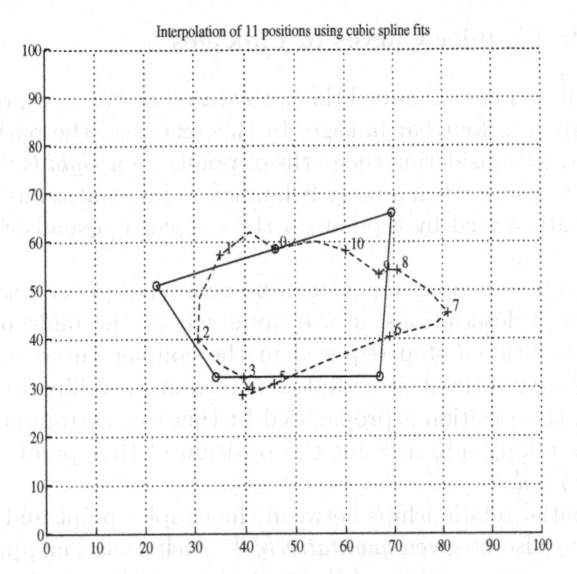

Fig. 6.7. Path connecting a sequence of *approximate positions of a mid-point* on the four-bar linkage (interpolation using cubic spline fits). The corresponding instantaneous positions of the mechanism are given in Figure 6.6.

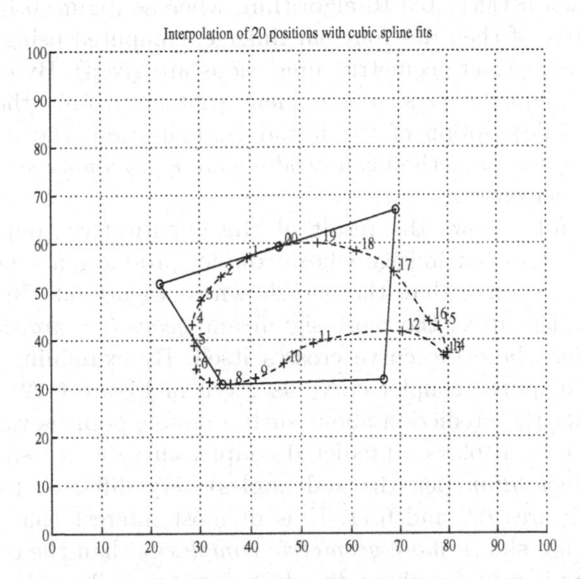

Fig. 6.8. The path of a mid-point on the floating link constructed by interpolating a sequence of 20 instantaneous positions through cubic spline fits. Based on the assumption that *the exact lengths of the linkage are given*, each instantaneous configuration is computed using Hunt's *algebraic* method [91].

Example 6.4: Coupler Curves in Linkages

Example 6.3 has examined the path traced by the mid-point on the floating link of a four-bar linkage. In this example, the path problem is extended by considering the paths of points on a *rigid body attached to the floating link*. Such a body is known, in kinematics, as a *coupler*. And the path traced by a point on the coupler is usually referred to as a *coupler curve*.

Approximate coupler curves can be generated given a set of qualitative spatial descriptions, in the same way as the mid-point curve. The only additional step required in the coupler curve case is that once a position of the floating link is determined, as illustrated in Example 6.4, this position is propagated further to a certain point in the coupler by taking into account the position of that point *relative to the floating link*.

The spatial relationships between the coupler point and the floating link may also be given *qualitatively*. In such a case, an approximate quantitative representation of their relative position can be generated using the ANNEAL-based local search, as discussed in Example 6.3.

Figure 6.9 presents a coupler curve of a four-bar linkage, generated using the C_INSTANT_CONFIG algorithm, whereas Figure 6.10 shows a coupler curve of the same four-bar linkage, computed using the algebraic method (exact geometric dimensions are given). By comparing these two coupler curves, it is evident that given only the qualitative spatial description of the initial configuration, the *qualitative-quantitative search* method can produce an *approximate sketch* of the actual coupler curve.

Figure 6.11 shows the result of the quantitative coupler curve search for a four-bar linkage whose coupler produces a topologically more complex curve than the one shown in Figure 6.9. In this predicted coupler curve, the most significant *geometric property* is the *double point* where the curve crosses itself. By examining the algebraically computed coupler curve as given in Figure 6.12, it can be realized that the prediction about such a double point is valid.

Figure 6.13 displays a predicted coupler curve which starts at an *initial configuration* (lengths and angles) very different from those given in Figures 6.9 and 6.11. It is of most interest that this coupler curve has shown *more geometric complexity* than the one in Figure 6.11, as it contains three double points (as well as three knots). If compared with the actual coupler curve generated using the algebraic method, as shown in Figure 6.14, it can be readily noted that the predicted curve has correctly captured the significant geometric properties of the actual curve.

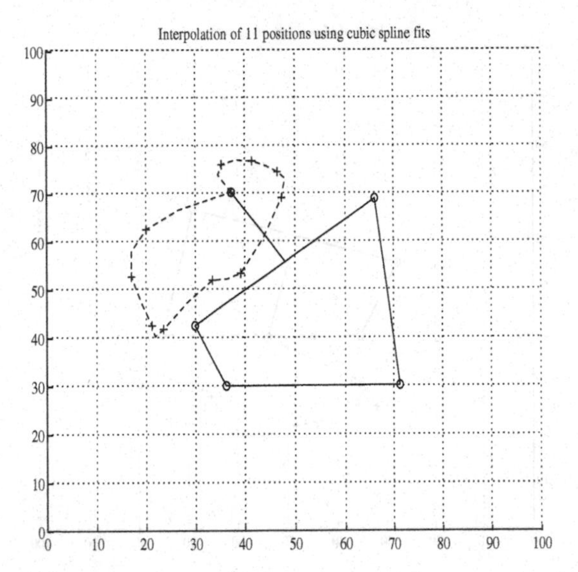

Fig. 6.9. A coupler curve derived directly from *qualitative spatial descriptions.* The *qualitative-quantitative search* method can produce an *approximate sketch* of the actual coupler curve, as shown in Figure 6.10.

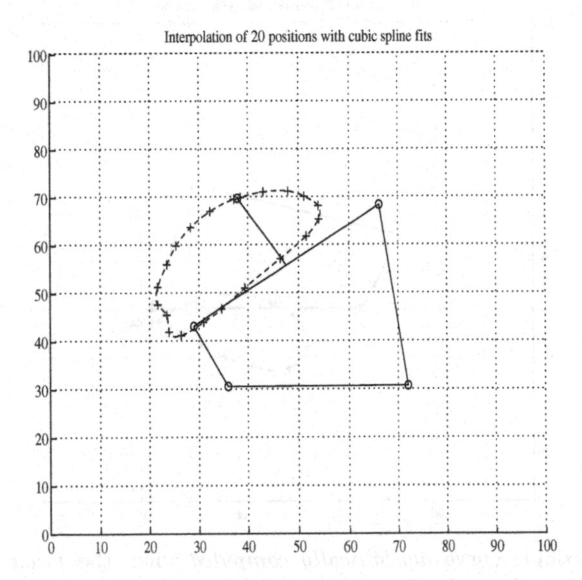

Fig. 6.10. A couple curve *algebraically computed* for the four-bar linkage, as shown in Figure 6.9. It is assumed that *the exact lengths of the linkage are given.*

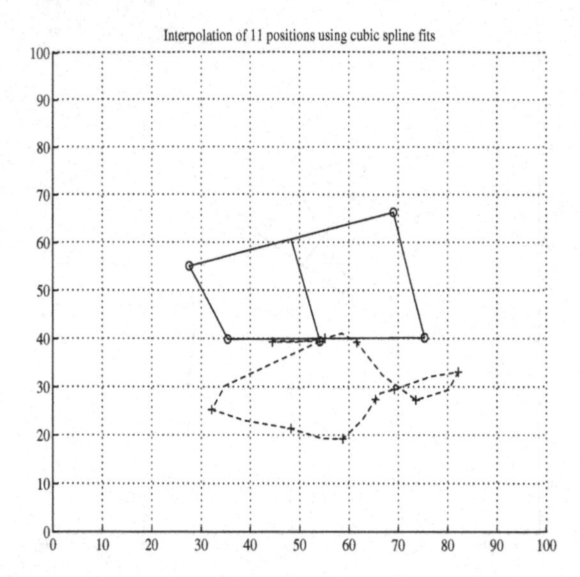

Fig. 6.11. A coupler curve derived from *qualitative spatial descriptions.* It contains a *double point* which correctly models the actual coupler curve, as in Figure 6.12.

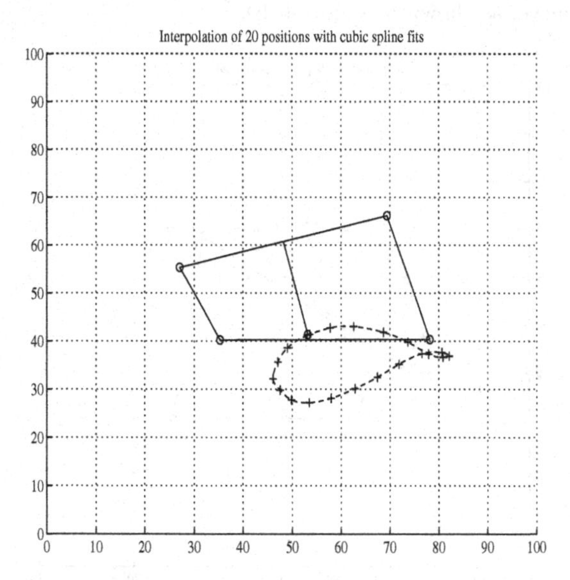

Fig. 6.12. A couple curve *algebraically computed* when the *exact lengths* of the four-bar linkage, as shown in Figure 6.11, are given.

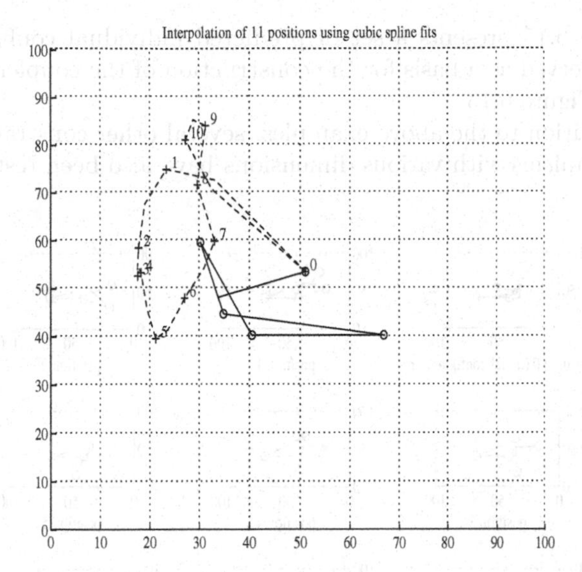

Fig. 6.13. Geometrically significant properties (i.e., *three double points* and *three knots*) are predicted in this coupler curve, even when the lengths of the linkage are *qualitatively* known. All these properties have correctly captured the ones in the actual coupler curve, as shown in Figure 6.14.

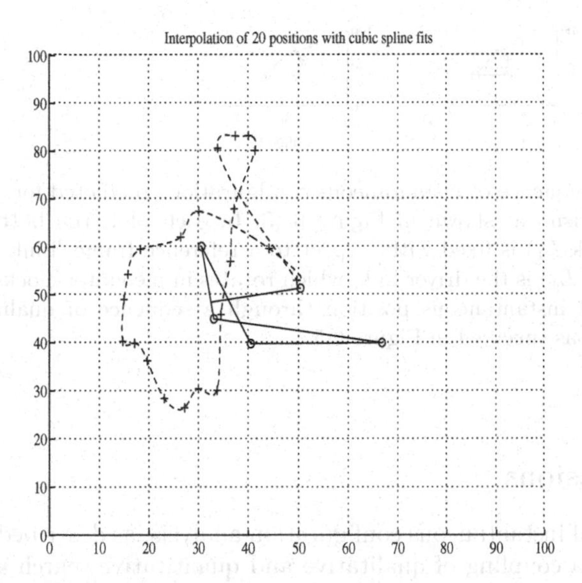

Fig. 6.14. An *algebraically computed* coupler curve given the *exact* lengths for the same four-bar linkage as in Figure 6.13.

Figure 6.15 presents a set of predicted individual configurations that has served as a basis for the construction of the coupler curve as given in Figure 6.13.

In addition to the above examples, several other constrained linkage mechanisms with various dimensions have also been tested.

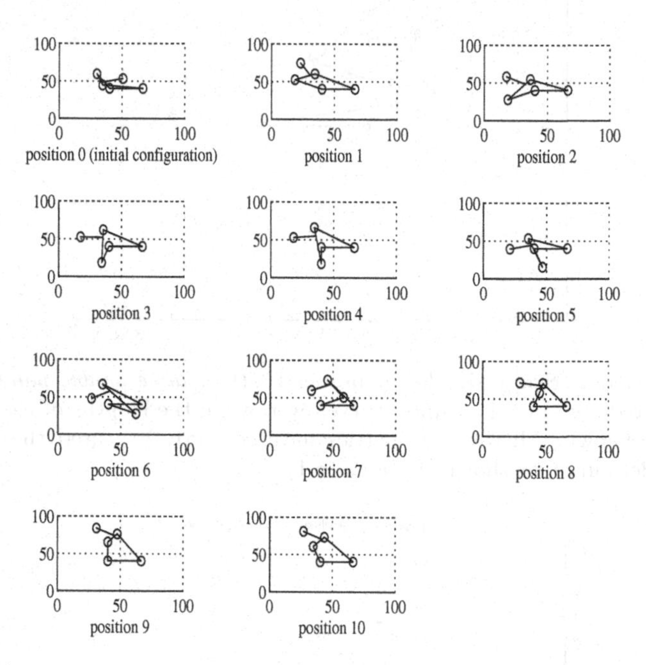

Fig. 6.15. A sequence of instantaneous configurations predicted for a *qualitatively-known* mechanism, as shown in Figure 6.13. In each plot, the bottom horizontal link (called link L_1) is fixed with respect to a reference frame. Link L_2 (numbered clockwise from L_1) is the driver link, which rotates in a counter-clockwise direction, from its initial instantaneous position through a sequence of qualitative angular displacements, as specified in Figure 6.5.

6.4 Discussions

The method of instantaneous configuration analysis, as described in this chapter, relies on a coupling of qualitative and quantitative search algorithms. It utilizes the qualitative configuration analysis algorithm (`QUALITATIVE_CONFIG`) to deduce a qualitative representation of an instantaneous configuration, and the quantitative configuration generation algorithm (`ANNEAL`) to yield a quantitative approximation of the qualitatively-predicted configuration.

This section discusses the features, advantages, and limitations of such an approach.

6.4.1 Features and Advantages

The major features of the current approach may be stated as follows:

1. **Correctness:** With respect to the qualitative configuration analysis, it is clear from Theorems 6 and 7 that if all the lengths of the links of a four-bar linkage mechanism (i.e., the basic component of complex linkages) and the position of its driver link are *qualitatively specified*, then the rest of the relative spatial relationships in the mechanism can be derived. Furthermore, the derived qualitative descriptions are *complete*. The completeness guarantees that some intervals mapped from the qualitative descriptions will contain the actual configuration parameters which may be found by quantitative configuration search.
2. **Complexity:** In the configuration analysis, a complex linkage mechanism is first decomposed into a set of independent sub-linkages (equivalent to four-bar linkages), as stated in the **Linkage decomposition** step of `C_INSTANT_CONFIG`. Subsequently, the propagation of configurations from one sub-linkage to its neighbors takes place when the quantitative configuration of the sub-linkage is obtained. In other words, the `C_INSTANT_CONFIG` algorithm does not directly propagate *qualitative spatial descriptions* across two sub-linkages, and hence it avoids the problem of a combinatorial explosion in the number of possible qualitative spatial inferences for the neighboring sub-linkages.
   With respect to the quantitative configuration search, a derived qualitative configuration description is used to place *limits over the search space* of quantitative configuration parameters (e.g., joint angles), and hence reduce the time complexity in the quantitative search.

The coupling of the qualitative and quantitative search techniques has also presented two advantages:

1. **Qualitative configuration handling:** When the exact geometry of a mechanism is not available, the current method, as illustrated in Examples 6.3 and 6.4, can yield an approximate path for a given point on a linkage. From such a representation, geometric properties of the characteristic path (e.g., double points, symmetry, loops, or peaks) as well as functions of the linkage (e.g., oscillator) can furthermore be identified.
2. **Non-algebraic formulation:** The method does not require special engineering of *algebraic formulas*. In the qualitative spatial inferencing, each step is controlled by a forward chaining algorithm which searchs a $\mathcal{QT}$ rule base, whereas in the simulated annealing, no formulation except the objective function is required. Therefore, this method is *less dependent* on the specific linkage problems.

6.4.2 Limitations

The proposed method, in general, has the limitation of inexactness. First, any two runs of the same process may generate two slightly different paths, both being an approximation of the actual path. Depending on the application, either approximate path may suffice to meet the need of mechanism analysis. Second, with different *annealing schedules*, the results of annealing can be different. However, to determine appropriate control temperatures is still an empirical matter that may be problem-sensitive.

Most of the constrained linkage mechanisms can be analyzed using the method described in this chapter. If an independent linkage contains a single slider, then the configuration in question becomes the sliding position with respect to a certain joint. However, if the linkage contains a higher pair, the configuration problems will become ill-defined in the proposed spatial representation framework. Hence, the current method does not cover such mechanisms.

6.5 Kinematic State Transitions in CSV Mechanisms

The methods presented in the preceding sections are in general limited to the qualitative kinematic analysis of contact-surface-invariant (CSI) mechanisms. This section addresses the problem of identifying kinematic state transitions of contact-surface-varying (CSV) mechanisms , in which the connectivity or contact-surface *varies* with the motion of mechanism bodies.

The function of a CSV mechanism such as a ratchet or an escapement mechanism can best be understood in terms of the interactions between individual bodies, i.e., the change of contacts or connectivity. This implies that in the qualitative analysis of a CSV's function, it will be desirable to consider each distinct contact as a discrete kinematic state and to describe the transition of the states by deriving a sequence of possible contact changes.

Stanfill [176] has developed a set of algebraic rules to determine where complex surfaces intersect and touch. In his Mack system, shapes are represented in terms of the sum and difference of primitive solids. The underlying objective is to model what seems intuitive and fundamental for humans to reason about the interactions of objects. However, since the mechanism parts considered are quite limited, this approach is unable to handle CSV mechanisms of complex shapes.

Faltings has proposed a method of qualitative spatial reasoning for analyzing kinematic state transitions [49]. This method, utilizing configuration-space representation, may be described as follows.

In Figure 6.16, ϕ and θ are used to describe the positions of bodies A and B, respectively. These two parameters constitute the generalized coordinates (i.e., independent coordinates) of the CSV mechanism. The space spanned by such parameters, which characterizes all positions of mechanism bodies, is called the mechanism's *configuration space* [143, 19, 138]. According to Faltings's method, if the configuration space and the constraints for its valid subspace (corresponding to all legal configurations) can be computed, then a set of possible "places", where bodies are in contacts can further be derived. Thus,

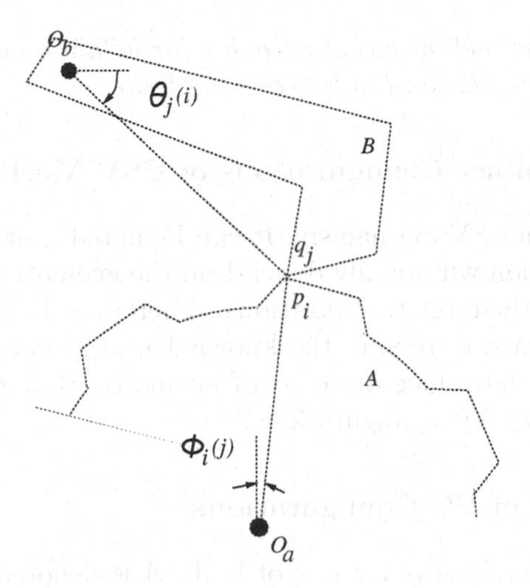

Fig. 6.16. A CSV mechanism.

by means of envisionment analysis, a sequence of qualitative state transitions can be inferred.

In the method to be described herein, what is required for deriving kinematic states and transitions comprises some specific configurations, i.e., *a subset of CS*, as well as the information about the placement of vertices of one body with respect to the edges of another in those configurations. In other words, the derivations depend only on incomplete configuration information and hence reduce the complexity involved in computing a full configuration space.

In this study, we consider primarily the type of CSV mechanism that is composed of two bodies and assume that (1) the boundary of each body can be described by a simple polygon (i.e., a set of connected non-crossing line segments) and (2) the degree of freedom for each body is equal to one. As for other CSV mechanisms, methods of analysis can be obtained by modifying those for the simpler, but similar, two-body mechanisms.

Here, a kinematic state is defined as a specific sliding contact between a boundary vertex and a boundary line segment by which a motion is transmitted. A state transition refers to the change of sliding contacts. A vertex-contact (VC) configuration refers to the configuration of a CSV mechanism in which two interacting bodies have a direct contact at their boundary vertices. The typical problem to be solved is stated as follows:

> *Given, in a two-body CSV mechanism, a set of VC configurations, and the vertex placement of the driving body with respect to the boundary of the driven body corresponding to the mechanism's VC configurations, determine the interactions of bodies, i.e., the changes of direct contacts as well as sliding motions between two bodies during a motion of the*

mechanism. For each new contact point (or within a kinematic state), find the velocity relationship between the bodies.

6.5.1 Vertex-Contact Configurations of CSV Mechanisms

Figure 6.16 shows a CSV mechanism. It can be noted that the change of contacts during a motion will mainly depend on the geometry of the mechanism bodies as well as their relative positions. Therefore, in order to determine kinematic states, we will require the knowledge of geometric configurations. In this section, we introduce a means of geometric characterization of CSV mechanisms utilizing VC configurations.

The Description of VC Configurations

In Figure 6.16, a series of n vertices of body A is denoted by p_i, where $1 \leq i \leq n$, and that of m vertices of body B is denoted by q_j, where $1 \leq j \leq m$. Suppose that distances $o_a p_i$, $o_b q_j$, and $o_a o_b$, and the lengths of boundary segments are given. Thus, the VC configurations of the mechanism composed of bodies A and B (in this case vertices p_i and q_j are in a direct contact) can be easily described by deriving (e.g., from *qualitative trigonometry*) $\phi_i(j)$ and $\theta_j(i)$. To be more general, the description of VC configurations can also be extended to cover prismatic-pairing bodies in CSV mechanisms. In such cases, the angles of each boundary segment relative to a fixed reference frame (rather than $o_a p_i$ and $o_b q_j$) must be given. Hence, the description of VC configurations comprises a set of positions in the sliding motion directions, as shown in Figure 6.17.

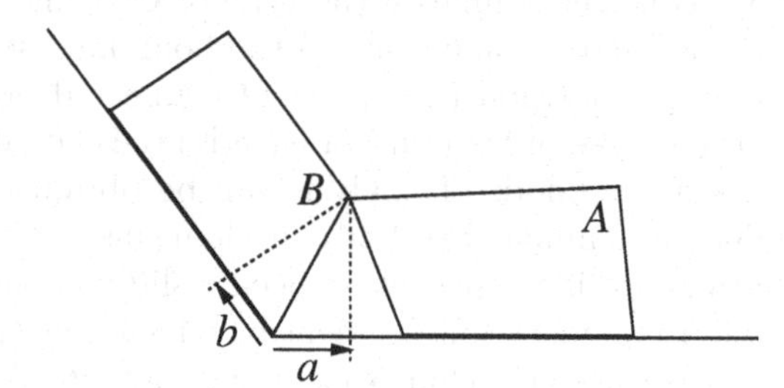

Fig. 6.17. The geometric description of prismatic-pairing bodies in a CSV mechanism.

6.5.2 Placement of Vertices in VC Configurations

In order to derive the transitions of kinematic states, more information than the description of VC configurations will be required. Suppose that a boundary vertex is in direct sliding contact with a line segment and at some point of the line segment the sliding motion *fails* to proceed due to certain geometric constraints. Note that, if after the *terminating contact point* the sliding motion was imaginarily continued, an overlap will occur between the boundaries of the bodies. Furthermore, if we let the contact point slide to the end-point of the boundary segment, then the overlapping will change accordingly into one of the three possible positions presented in Figure 6.18.

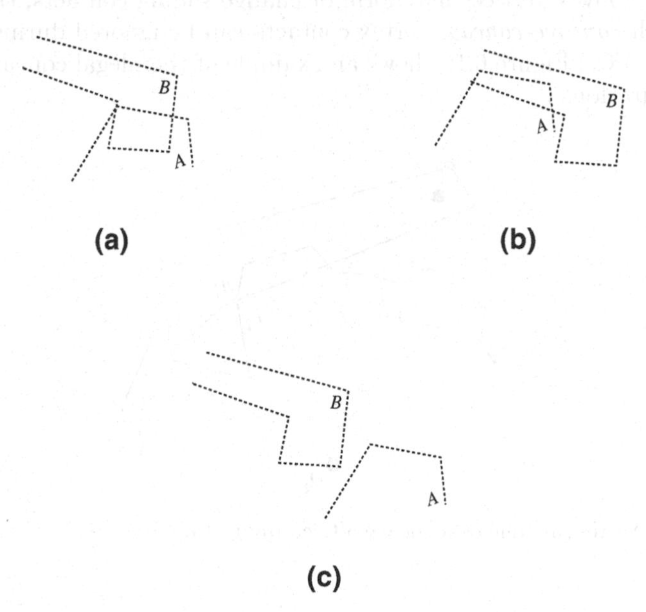

Fig. 6.18. Three possible positions into which the overlapping between mechanism bodies may change.

From Figure 6.18, it can be observed that each position corresponds to a change in relative vertex placement as the sliding motion is continued from a given contact point, for instance an initial contact point or a vertex-contact point, to an end-point of the boundary line segment. Here, vertex placement refers to the placement of the vertices of one body relative to the boundary line segments of another in each VC configuration. Such an observation implies that the potential overlap(s) can be determined simply by comparing the placements of vertices in two consecutive VC configurations, i.e., the VC configurations corresponding to two consecutive vertex contacts at which the imaginary sliding motion starts and ends. And consequently, from the information of potential overlap(s), a possible contact point between two bodies can be detected.

With respect to the example in Figure 6.16, the vertex placements may be computed as follows: for a vertex on body A, p_i, in direct contact with another vertex on body B, q_j, find the placement of the rest of A's vertices with respect to all of B's line segments. Repeat this process for all i's and

j's. In order to keep track of the relative vertex placement, the direction of each polygonal curve should be defined. Thus, given the direction of a certain boundary line segment, the placement of a vertex can be described in terms of which side of the extended line it lies on. This may be done, for instance, by evaluating the vertex in the equation of a line segment. The vertex placement in the case of prismatic-pairing bodies can be derived similarly.

As only convex vertices may form or change sliding contacts, the configurations with *concave-concave* vertex contacts can be ignored during the computation of VCs. Figure 6.19 shows an example of the illegal concave-concave VC configurations.

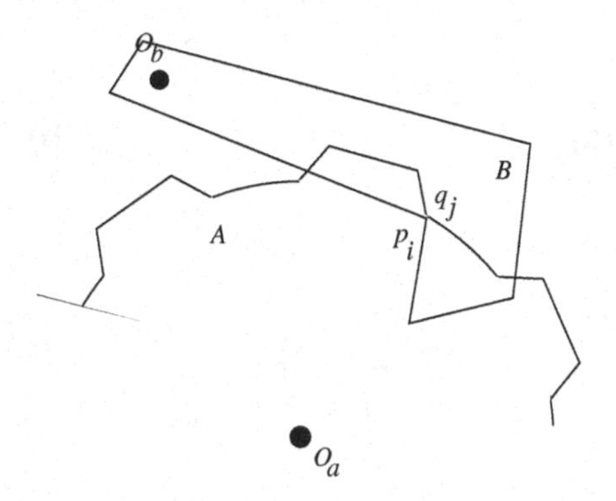

Fig. 6.19. An illegal concave-concave VC configuration.

6.5.3 Identification of Kinematic State Transitions

This section describes how the above-mentioned representation of VC configurations is used in reasoning about the motions of CSV mechanisms.

Change of Sliding Contacts

The following is an algorithm for identifying a sequence of contact changes in the motion of a CSV mechanism:

Algorithm SLIDING_CONTACT

Input: The description of VC configurations with respect to a fixed frame and the corresponding vertex placements of a driving body, A, relative to each extended boundary line segment in a driven body, B. Note that, in the initial configuration, the contact may exist between a boundary line segment and a boundary vertex.

Output: A sequence of sliding contact changes caused by a given driving motion.

1. Start with the initial configuration, keep track of the coordinate values and the vertex placement. Assign them to a coordinate vector (a, b) and a placement vector *previous_vposition*, respectively. Suppose that the sliding vertex moves to one end-point of the line segment, and find the corresponding VC description.
2. If the direction of A's position change is inconsistent with that of the given motion, then go to Step 1, else keep track of the VC description and the vertex placement, and assign them to a coordinate vector (a', b') and a placement vector *current_vposition*, respectively.
3. Identify the possible overlap(s) during the sliding motion by comparing *previous_vposition* with *current_vposition*:
   a) If the position of a vertex of A relative to an extended boundary line of B changes from outside to inside (note that, if we walk counter-clockwise along the boundary of B, the right hand side is defined as the outside), then there exists a possible overlap, and the vertex is considered as a potential sliding contact point.
   b) If from *current_vposition* it is found that two connected vertices of A have different placements with respect to at least two boundary line segments of B (note that a contact vertex is considered to be outside of the contact line segment), then there exists a possible overlap. Either one of these two vertices will be the new sliding contact point, or the boundary segment formed by the two vertices will be in direct sliding contact with a vertex in B.
   c) If there no possible overlap region is detected, then conclude that the previous sliding motion is valid and go to Step 5.
4. For each possible contact vertex p_i, start with the *furthest* vertex in terms of its position in A and find the line segment to which p_i is an inside vertex.
   a) Compute the VC configuration, formed by p_i and one of the line segment end-points. The end-point is chosen such that p_i has a *larger* position. Update (a', b') and *current_vposition*. Compare a' with a; if the result is consistent with the direction of the given driving motion and if there cannot be found any possible overlaps in *current_vposition*, then conclude that the next sliding contact is *between the line segment and* p_i and go to Step 5, else go to Step 4c.
   b) Compute the VC configuration, formed by p_i and one of the line segment end-points. The end-point is chosen such that p_i has a *smaller* position. Update (a', b') and *current_vposition*. Compare a' with a, if the result is consistent with the direction of the given driving motion and if there cannot be found any possible overlaps in *current_vposition*, then conclude that

the next sliding contact is *between this end-point of the line segment and the line segment formed by p_i and its succeeding vertex* and go to Step 5, else go to Step 4c.
   c) If there are *new* possible contact vertices, then recursively run Step 4.
5. Keep track of the sliding contact change. Test whether a blocking configuration has been reached. If not, then consider the VC configuration as an 'initial' configuration (i.e., assign (a', b') to (a, b) and *current_vposition* to *previous_vposition*). With the same sliding vertex, go back to Step 1.

This algorithm can be used to identify a sequence of sliding contact changes between two bodies of a CSV mechanism if it is given the description of VC configurations and the corresponding vertex placement information. Figure 6.20a depicts the initial configuration of a CSV mechanism. It can be verified that by computing its VC configurations and applying the contact analysis algorithm, we can derive the sliding contact changes (i.e., contact points), as shown in Figure 6.20.

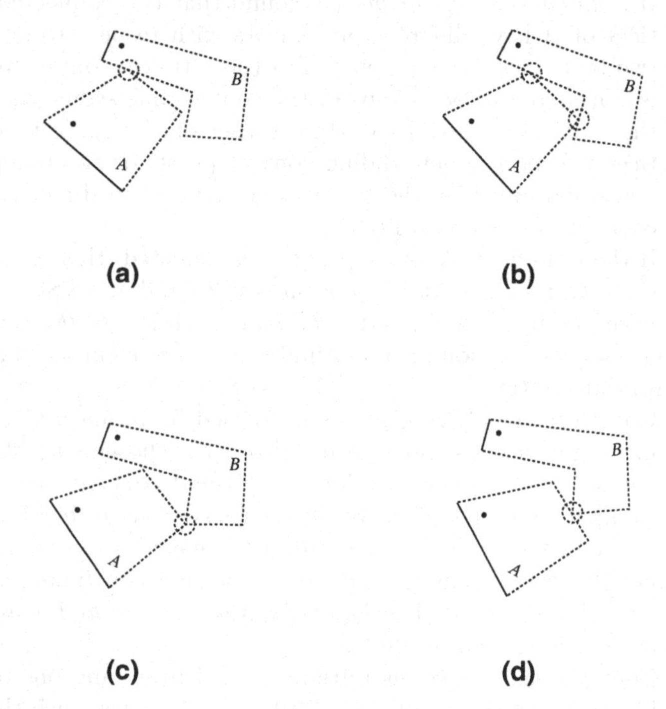

Fig. 6.20. The sliding contact changes between two bodies of a CSV mechanism.

However, it may be noticed from the above algorithm that the detail of Step 5 has been left out. Let use now discuss how the test of mechanism movability is performed. First, we consider the following theorem:

Theorem 12 (Motion transmission by direct contact (see Figure 6.21)). *Suppose that a driving body A is in direct contact with a driven piece B. If, with respect to a reference frame, the directions of the potential motion vectors of both A and B are not perpendicular to the common normal of the contact surface and both motion vectors are on the same side of the common tangent, then the motion can be transmitted from A to B. The partial ordering relationship between their absolute velocities can be determined from the angles formed by each motion vector and the normal.*

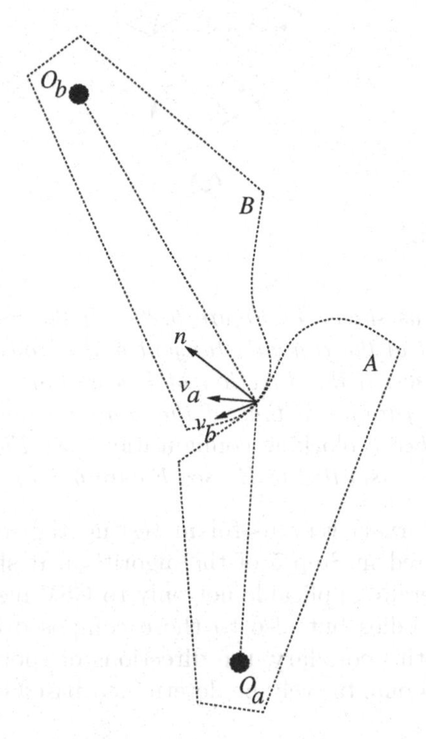

Fig. 6.21. Transmission of motion by direct contact.

From Theorem 12, we can further derive the following corollary:

Corollary 1 (see Figure 6.22). *Let a driving body A and a driven body B be in a direct contact. With respect to a fixed reference frame, if the motion direction of A is the same as that of the common tangent between A and B (i.e., parallel to the current contact boundary line segment), then A will be in sliding motion relative to B and no motion will be transmitted to B, i.e., B*

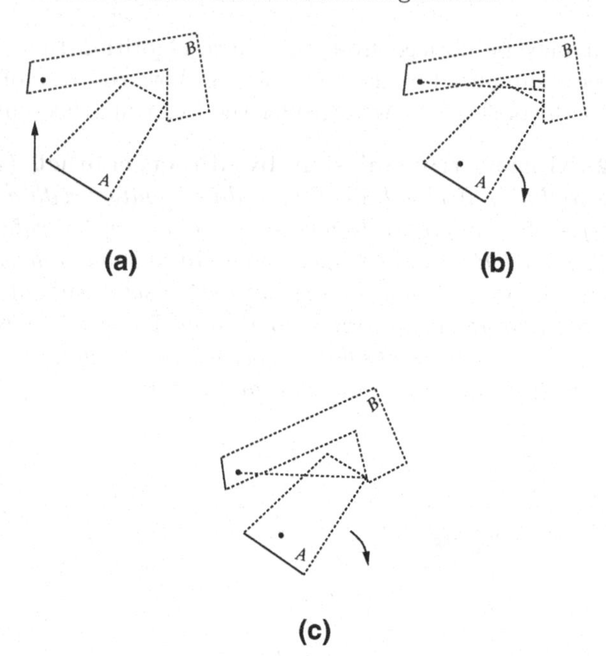

Fig. 6.22. Corollary 1.

remains stationary, as shown in Figure 6.22a. If the motion direction of A is different from that of the common tangent and is toward B, then we have the following two cases: if the direction of B's motion vector (i.e., potential motion direction) is parallel to that of the common tangent, then the CSV mechanism has reached a blocking configuration *(see Figure 6.22b); else the motion of A will be transmitted to B (see Figure 6.22c).*

The above corollary is very useful in testing the movability of a CSV mechanism, as required in Step 5 of the algorithm. It should be noted that this corollary is generally applicable not only to CSV mechanisms composed of revolute-pairing bodies but also to those composed of prismatic-pairing bodies. In applying this corollary, the directions of (potential) motions and common tangent can qualitatively be determined based on the description of a CSV mechanism.

Velocity Relationship between Two Bodies

In this subsection, we discuss how to determine the change of motion transmission between two bodies of a CSV mechanism. Suppose that at each new sliding contact inferred using the preceding algorithm, the motion transmission relation is approximately constant. Thus, if we can find the *unique* velocity relationship between two bodies within each sliding motion, we will be able to obtain the change of motion transmissions, corresponding to the change of sliding contacts. Hence, we will completely solve the problem of deriving kinematic state transitions of a CSV mechanism.

The idea behind velocity relationship analysis is straightforward. From the preceding subsection, we know that given the descriptions of a CSV mechanism and its VC configurations, we can find, by means of simple calculations, the changes in (angular) position of two bodies as a contact vertex slides along a boundary segment from one end-point to another. As a result, we can determine the angular velocity relationship between the two bodies during sliding motion. And since the distances from the origin of each coordinate to the vertices of a body are known, we can also derive the corresponding linear velocity relationship. If all these mentioned quantities are given in terms of qualitative values or relations, the qualitative description of the velocity relationship can therefore be obtained.

In order to illustrate the method of velocity relationship analysis, we will consider an example of sliding contact change, as shown in Figure 6.20. It should be noted however that this method is not restricted to revolute-pairing CSV mechanisms.

Let us denote, with respect to the local reference frames of individual bodies in the configuration where vertices p_i and q_j are in a direct contact, the vertices of bodies A and B as p_i and q_j, respectively, and the angular positions of $o_a p_i$ and $o_b q_j$ as $\phi_i(j)$ and $\theta_j(i)$, respectively (see Figure 6.23). Suppose that the angular velocity of body A is given, and we want to find the linear velocity relationship between the two bodies in the *second* sliding contact.

By applying the contact analysis algorithm, we know that the second sliding contact must be formed by vertex p_2 and boundary segment $q_1 q_2$. Therefore, we compute the coordinate value differences of $\triangle\phi$ and $\triangle\theta$ as p_2 slides from q_1 to q_2, i.e., $\triangle\phi = \phi_2(2) - \phi_2(1)$ and $\triangle\theta = \theta_2(2) - \theta_1(2)$. From these angular differences, we find the angular velocity relationship between the two bodies (as the motion transmission relation is here approximated to be constant). More specifically, we describe the motion transmission relation for the second sliding contact as a ratio between two angular differences, i.e., $\omega_a/\omega_b = \triangle\phi/\triangle\theta$. If the distances from the coordinate origin in body B to vertices q_1 and q_2 are close enough, we can further determine the approximate relationship between the magnitudes of two linear velocities at the sliding contact point by $v_a/v_b = (\triangle\phi \times o_a p_2)/(\triangle\theta \times o_b q_2)$.

6.6 Summary

In this chapter, we have discussed how to model and reason about the instantaneous configurations of a contact-surface-invariant (CSI) mechanism. Our general approach is composed of two phases. The first phase is to reason about a qualitative configuration using the incomplete dimensional information about a mechanism. The second phase is to further generate an exact

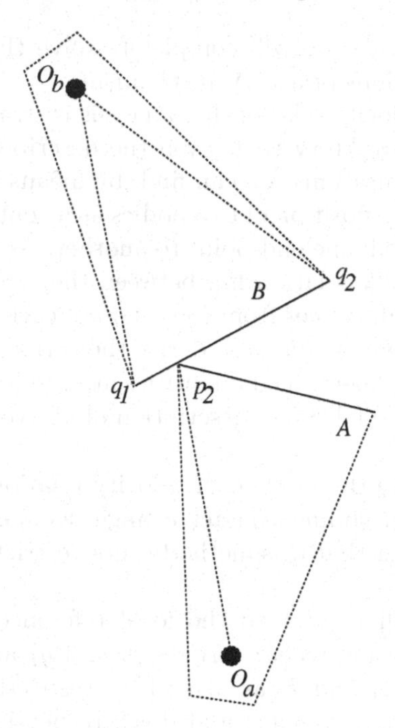

Fig. 6.23. The velocity relationship between two interacting bodies.

configuration based on the qualitative result from the first phase. We have provided detailed algorithms for both.

In addition, we have also addressed the problem of contact-surface-varying (CSV) mechanisms by identifying kinematic state transitions based on VC configurations and the vertex placements with respect to such configurations.

We have demonstrated the usefulness of qualitative mechanism reasoning with linkage and two-body CSV examples.

7

How to Reason about Velocity Relationships

In constrained linkage mechanism analysis, apart from knowing configurations and trajectories, it is desirable to know the velocities of various links as well as their interrelationships given certain kinematic constraints. This chapter introduces a method for deriving instantaneous velocity relationships among constrained bodies of a mechanism. The method utilizes the qualitative kinematic properties (i.e., *instantaneous rotation center*) of mechanisms, and permits computationally efficient solution to the problem of deriving velocity relationships.

Liu [132] previously proposed a qualitative approach to velocity analysis based on instantaneous rotation centers. His approach relied on a set of naive spatial inference rules and generated spatial envisionments which were sometimes considered too ambiguous to be practically useful. The current work, however, derives instantaneous centers by means of the qualitative-quantitative configuration analysis as presented in Chapter 6. Since the qualitative configuration modeling step is based on a complete set of trigonometric rules and of reasonable precision, the results of spatial analysis will contain less uncertainty. Furthermore, since quantitative spatial relationships are generated for each instantaneous configuration, successive qualitative modeling of instantaneous rotation centers will avoid the problem of combinatorial complexity caused by the ambiguous envisionments.

7.1 Instantaneous Rotation Center

Before describing the qualitative approach to velocity analysis, it will be useful to recall one of the properties of an *instantaneous rotation center* (instantaneous center) [91], that is,

> *The instantaneous linear velocities of points on a given link are perpendicular to the lines joining these points with an instantaneous center.*

Based on the **V-direction axiom**, the instantaneous center of an individual link in a linkage can be located. Consider the four-bar linkage, as shown in Figure 7.1. It can be readily realized that point D is a point on both link L_3 and link L_4. As a point on L_4, D moves with respect to L_1 about the center A and thus in a direction normal to L_4 itself. The motion of D on L_3 with respect to L_1 is in a direction perpendicular to L_4. Hence, by definition, the instantaneous center will lie on the line through D and the direction of L_4. Similar reasoning can be applied to infer that the instantaneous center of L_3 must also lie on the line through point C normal to L_2. Hence, P_{13} is at the intersection of two lines extended from L_2 and L_4.

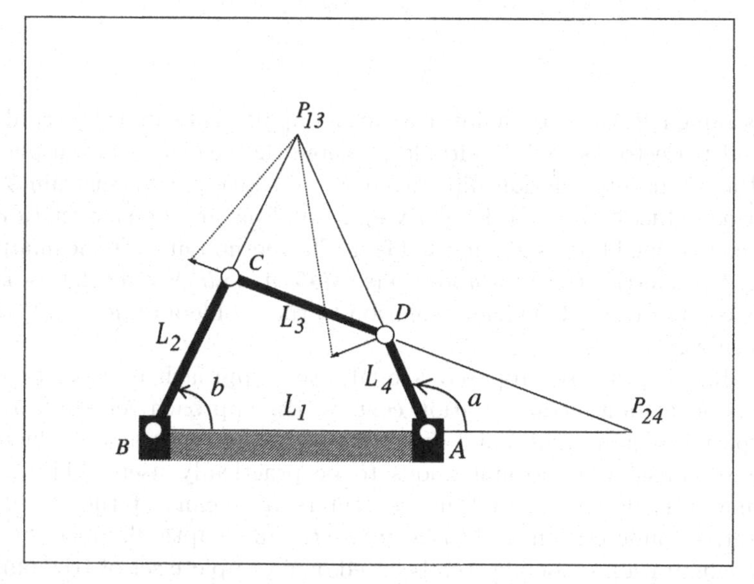

Fig. 7.1. An illustration of *instantaneous rotation centers.* By means of building instantaneous centers, the problem of qualitatively describing velocity relationships can be reduced to that of qualitative spatial analysis.

The concept of instantaneous centers is essential to motion analysis. It provides a geometric method in determining the relationship between two linear velocities of the same mechanism. The relative instantaneous velocities have the following geometric property:

> *The instantaneous linear velocity of a point on a given link is proportional to its radius of instantaneous rotation.*

Based on the **V-magnitude axiom**, it is possible to determine the *velocity distribution* on each link, and to infer the *motion transferred* from one link to another.

It is important to point out that the notion of instantaneous rotation can be applied not only to derive the linear velocity relationships, but also to infer angular velocity relationships. As an example, shown in Figure 7.1, the instantaneous center for the relative motion of links L_2 and L_4, namely P_{24}, is found at the intersection of two lines extended from L_1 and L_3. The two links behave instantaneously as though they were *spur gears* having internal contact at P_{24}. Hence, their angular velocity ratio is given by

$$t_{24} = \frac{\omega_a}{\omega_b} = \frac{BP_{24}}{AP_{24}} \tag{7.1}$$

which provides a simple and quick method of finding the instantaneous angular velocity relationships.

The idea of instantaneous centers has in fact been employed in classical kinematics for analyzing velocities. However, such an analysis is mainly based on the graphical construction of all the centers and the visual determination of the distances from a given center to certain points on the corresponding link. In order to locate the centers, it usually requires the use of Kennedy's Theorem [91], which states that *the centers of any three planar bodies lie on a straight line.*

In the method presented below, the same concept of instantaneous centers is used. Instead of graphically analyzing a linkage and constructing its centers, this method applies the qualitative-quantitative configuration analysis algorithm, as presented in Section 6.1, to infer the approximate locations of the centers. The location of an instantaneous center is determined according to axioms of instantaneous velocity direction and magnitude (i.e., **V-direction** and **V-magnitude**, respectively). The distances from a given center to other points of interest are first *qualitatively* inferred using algorithm `QUALITATIVE_CONFIG`, and then *quantitatively* located using algorithm `ANNEAL` (the simulated annealing). Thereafter, the analysis of instantaneous velocity relationships in the linkage is carried out.

7.2 Velocity Relationship Analysis

This section discusses how to reason about the transfer of motion between two links as well as the velocity distribution on a given link. The qualitative-quantitative analysis of velocities will be illustrated with linkage examples.

Definition 19 (Velocity distribution and motion transfer). *Given any instantaneous configuration of the linkage, the velocity distribution in a certain link is defined as the absolute linear velocities of a set of points on the link. The absolute linear velocity relationship between a driver link and a driven link is referred to as motion transfer.*

In general, there are two primary types of motion transfer problems to be considered, with respect to whether the desired velocity is on a *floating* link or on a *follower* link. They are denoted as *TransFlt* and *TransFlw* types, respectively. The only distinction between the follower link and the floating link is that the former has one fixed end-point (i.e., connected to a fixed link where the frame of reference is located), whereas the latter has no fixed end-point. Under each of these two types, there exist two situations to be distinguished, depending on whether the input and the desired velocities are located in a single four-bar (or equivalent) linkage, or in two different four-bar linkages. These situations will be denoted by subscripts "*within*" and "*between*", respectively. Hence, in total, the analysis method will deal with four specific types of problems, as illustrated in Figure 7.2.

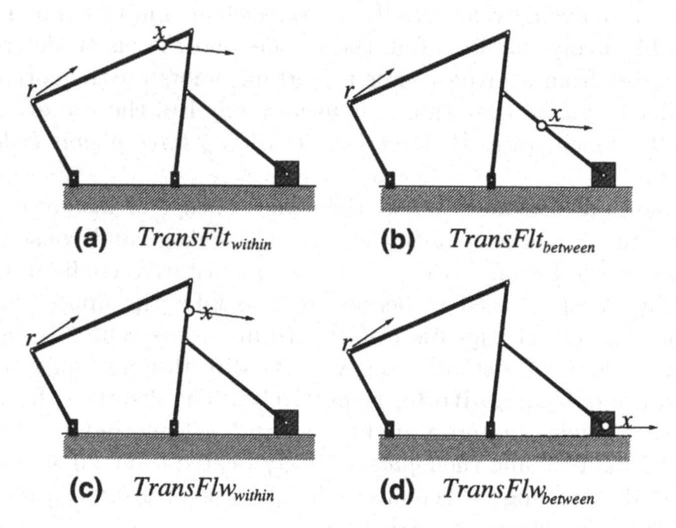

Fig. 7.2. Taking the linkage mechanism of Figure 6.4 for example, four types of qualitative velocity distribution problems can be identified, depending on whether or not the desired velocity (at point x) is on a *floating* link, and also whether or not the input velocity (at point r) and the desired velocity (at point x) are located in a single four-bar (or equivalent) linkage.

The details of the method are given as follows in an algorithm for analyzing motion transfer (C_VELOCITY). It should be noted that when the velocities of two or more points on a given link are analyzed, the *velocity distribution* of this link is obtained. Hence, the velocity distribution problem can be viewed as a special case of the motion transfer problem.

Algorithm C_VELOCITY

Input: Link lengths and a driver joint angle in a linkage mechanism (qualitative or quantitative), linear velocity of a specific link, and point(s) whose relative velocity is of interest.
Output: Quantitative velocity relationships.

1. **Linkage decomposition:** Find a set of *independent* sub-linkages (equivalent to four-bar linkage mechanisms) from the given linkage.
2. **Velocity input:** If the input linear velocity is not located at an end-point, compare distances from the fixed end-point of the link to that location, and to another end-point.
3. **Instantaneous center:** Start with the sub-linkage containing a given driver link. Locate the instantaneous center of its floating link by applying algorithm I_CENTER (see below).
4. If it is a $TransFlt_{within}$ problem, or a $Trans***_{between}$ problem and the linkage shares its floating link with another sub-linkage,
   a) Compare distances from the fixed end-point to the driver link and to the floating link.
   b) If it is a $TransFlt_{within}$ problem, based on obtained distance ordering, derive the velocity relationship between the known driver link and the desired point, and exit; else go to Step 6.
5. If it is a $TransFlw_{within}$ problem, or a $Trans***_{between}$ problem and the linkage shares its follower link with another sub-linkage,
   a) Compare distances from the fixed end-point to the driver link and to the follower link. If the desired point or the shared axis is not an end-point, further compare distances from the fixed end-point to the free end-point and to the desired point.
   b) If it is a $TransFlw_{within}$ problem, based on distance ordering obtained, derive the velocity relationship between the driver link and the desired point, and exit.
6. If the relative velocity at the *final* desired point is found, then exit; else find an associated sub-linkage and go to Step 2.

Symbol *** denotes either *Flt* or *Flw*.

The following algorithm provides details on how to locate an instantaneous axis for a floating link, as illustrated in Figure 7.3.

Algorithm I_CENTER

Input: Link lengths and a driver joint angle of a four-bar linkage (or equivalent) mechanism (qualitative or quantitative).
Output: The quantitative location of an instantaneous rotation center O of the floating link with respect to the fixed link L_1.

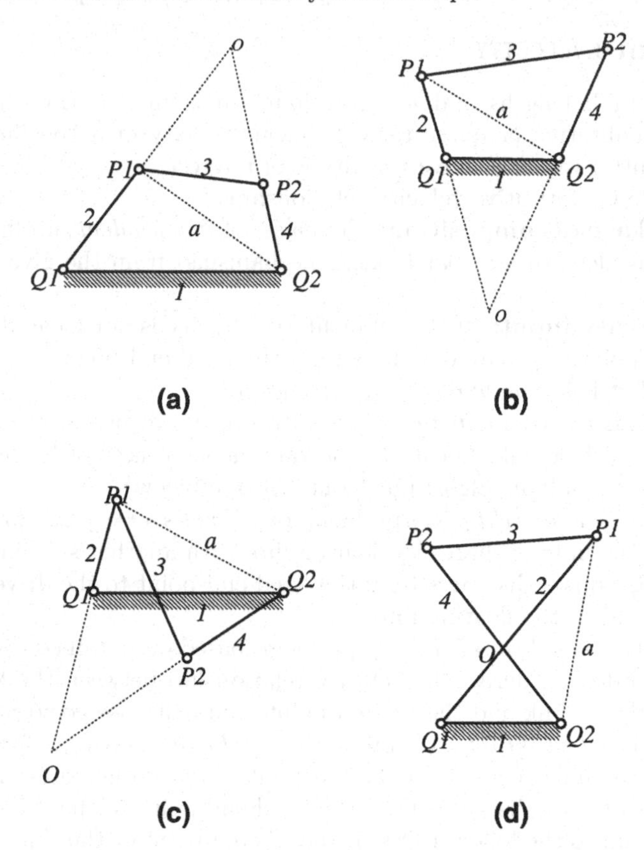

Fig. 7.3. Given a specific instantaneous configuration of a linkage, algorithm I_CENTER can generate spatial descriptions of the corresponding instantaneous center for a floating link.

1. Determine the joint angles from an intermediate link L_a (connecting L_1 and L_2) to L_2 and L_3.
2. **IF** $\texttt{MAX}(\theta_{2a}, \theta_{3a}) < \theta_{23}$ **THEN**
   a) **IF** $\texttt{SUM}(\theta_{12}, \theta_{14}) < \pi$ $\texttt{OR}$ $\texttt{SUM}(\theta_{23}, \theta_{34}) > \pi$ **THEN**
      $OQ_1 = \texttt{SUM}(OP_1, L_2)$ and $OQ_2 = \texttt{SUM}(OP_2, L_4)$ (see Figure 7.3a), where OP_1 and OP_2 with respect to L_3 are computed from $\texttt{DIFF}(\pi, \theta_{23})$ and $\texttt{DIFF}(\pi, \theta_{34})$.
   b) **ELSE**
      $OQ_1 = \texttt{DIFF}(OP_1, L_2)$ and $OQ_2 = \texttt{DIFF}(OP_4, L_4)$ (see Figure 7.3b), where OP_1 and OP_2 with respect to L_3 are computed from θ_{23} and θ_{34}.
3. **IF** $\texttt{MAX}(\theta_{2a}, \theta_{3a}) > \theta_{23}$ **THEN**
   a) **IF** $\theta_{2a} > \theta_{3a}$ **THEN**
      $OQ_1 = \texttt{DIFF}(OP_1, L_2)$ and $OQ_2 = \texttt{SUM}(OP_2, L_4)$ (see Fig-

ure 7.3c), where OP_1 and OP_2 with respect to L_3 are computed from DIFF(π, θ_{34}) and θ_{23}.

b) **ELSE**
OQ_1 = DIFF(L_2, OP_1) and OQ_2 = DIFF(L_4, OP_2) (see Figure 7.3d), where OP_1 and OP_2 with respect to L_3 are computed from θ_{34} and θ_{23}.

θ_{ij} denotes the joint angle between links L_i and L_j. The lengths and joint angles are determined using the QUALITATIVE_CONFIG and ANNEAL algorithms. SUM(∗, ∗), DIFF(∗, ∗), MAX(∗, ∗), and MIN(∗, ∗) denote the sum, difference, maximum, and minimum of the two given parameters, respectively.

Figure 7.4 presents a schematic review of the method for deriving instantaneous velocity relationships in a mechanism.

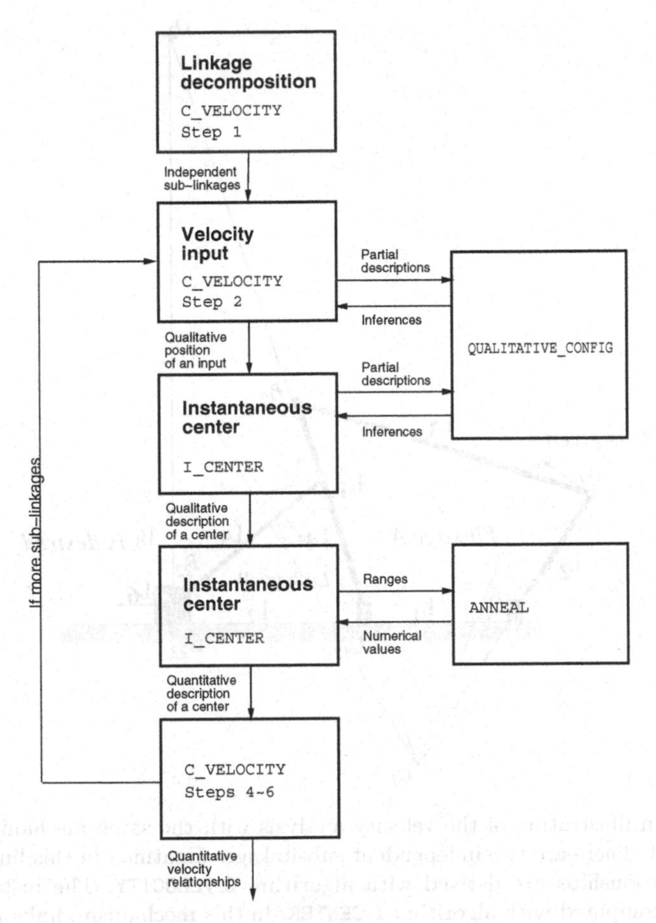

Fig. 7.4. A schematic review of the method for deriving instantaneous velocity relationships.

7.3 Examples

This section illustrates how algorithm C_VELOCITY can be applied to the qualitative velocity analysis, using the same linkage example shown in Figure 6.4.

Example 7.1: Velocity Relationships in a Four-Bar Linkage

Consider the mechanism shown in Figure 6.4, the motion transferred from an input crank (L_2) to a slider (L_6) is to be analyzed, as indicated in Figure 7.5. By applying graph searching, two independent four-bar linkages, A and B, can be found, and further, by definition, it is known that the problem is of $TransFlw_{between}$ type.

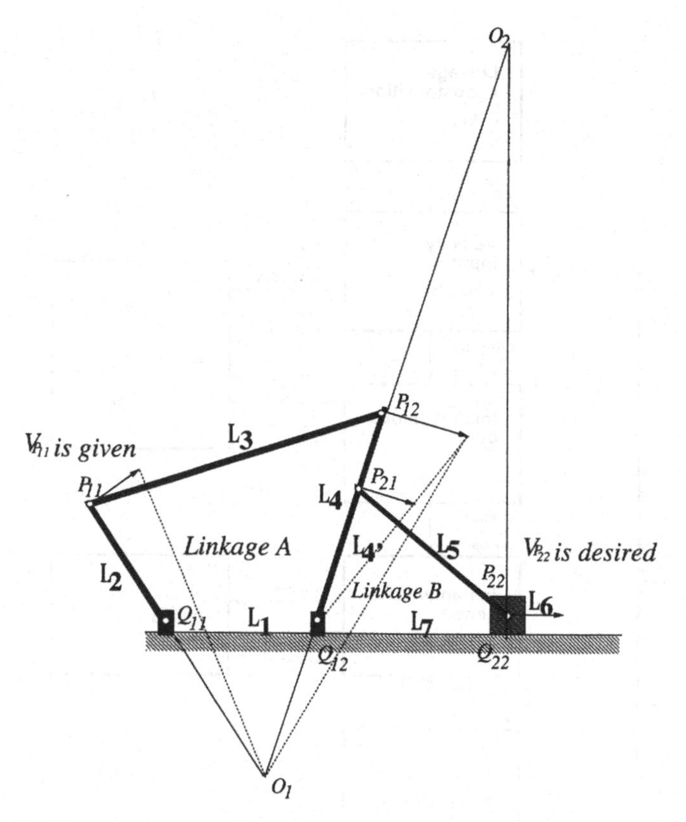

Fig. 7.5. An illustration of the velocity analysis with the same mechanism shown in Figure 6.4. There are two independent sub-linkages identified in this linkage. The velocity relationships are derived with algorithm C_VELOCITY. The instantaneous centers are computed with algorithm I_CENTER. In this mechanism, links L_1 and L_7 are both fixed with respect to a frame of reference. All the kinematic joints, except the joint between links L_6 and L_7 (a sliding joint), are revolute joints.

The velocity analysis starts with the linkage containing the driver link, i.e., linkage A. As the location of the linear velocity on the driver link is at its end-point, Step 2 of C_VELOCITY is bypassed. Next, the instantaneous rotation center of the floating link in linkage A is determined. In doing so, the above-mentioned I_CENTER algorithm is applied to obtain O_1Q_{11} and O_1Q_{12}, with respect to L_1, from θ_{12} and θ_{41}, and O_1P_{11} and O_1P_{12}, with respect to L_3, from θ_{23} and θ_{34}. From the results of O_1P_{11} and O_1P_{12}, the velocity relationship between $V_{P_{11}}$ and $V_{P_{12}}$ can be derived. Having computed the velocity at the joint of L_3 and L_4 with respect to the axis of L_4, it is possible to further analyze the velocity $V_{P_{21}}$ at the shared joint P_{21}.

Next, the second linkage, B, with the shared link as its driver link is considered, and the previous steps are repeated. Note that *the instantaneous center of the slider is located at infinity.* Thus, by applying I_CENTER, O_2P_{22} and O_2P_{21} can be derived, and consequently the relationship between $V_{P_{22}}$ and $V_{P_{21}}$. If all the velocity relationships obtained are combined, an approximate quantitative description of the motion transfer from L_2 to L_6, i.e., a relationship between $V_{P_{11}}$ and $V_{P_{22}}$, will be obtained.

Example 7.2: Velocity Constraints in an Ellipsograph Mechanism

Example 7.2 is concerned with the velocity relationships in an ellipsograph mechanism, as shown in Figure 7.6. Suppose that O is motionless with respect to a global fixed frame of reference. The constraint at point A allows the floating link ABP to translate along the y-axis and to rotate about the z-axis, while the constraint at point B allows ABP to translate along the x-axis and to rotate about the z-axis. The velocity at point P of the floating link relative to the velocity at A is desired.

In order to derive the velocity at point P, the intersection of L_A and L_B (i.e., the instantaneous center of the link ABP) is first located, which is denoted by I. Then, from the given position, $\angle\Phi$, of the link, the relationships between lengths AI and BI and between V_A and V_B are inferred. Similarly, the relationship between V_P and V_A can be obtained. Furthermore, the direction of V_P is known from $\angle AIP$. If V_P is expressed by a motion vector, then the approximate values of the velocity vector components in the x and y directions can be derived.

As the link moves, it may be desirable to know how V_P should change correspondingly in order to maintain V_A. Based on the above reasoning, the change can be readily analyzed. More specifically, with respect to the triangle $\triangle AIP$, if the angle $\angle PAI$ increases by a certain value, $\angle AIP$ and the distances AI and IP will change accordingly. As a result, a new velocity V_P can be determined.

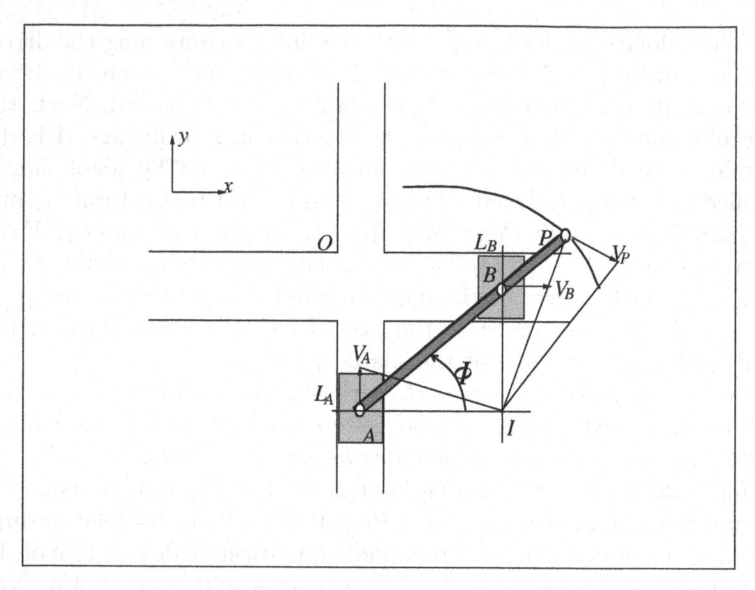

Fig. 7.6. Velocity analysis for an *ellipsograph* mechanism. In the mechanism, A translates along the y-axis and B translates along the x-axis. The velocity at point P, V_P, *relative* to the velocity at A, V_A, is desired.

7.4 Notes on the Application of *Velocity Analysis*

As a robotic application, the velocity analysis method illustrated in Example 7.2 can be used to *identify* external velocity constraints in a robot's task environment. Such velocity constraints are taken into account in programming the robot's compliant motions for certain manipulation tasks (such as turning the crank of the mechanism, as shown in Figure 7.6) [94, 189].

Mason [147] has proposed a method of planning robot compliant motions. This method requires that velocity constraints of individual links be propagated to a common point at the robot's effector by means of linear translational and rotational transformations. The new constraints obtained can then be translated into specific control strategies. This method outputs *accurate* velocity constraint information with respect to given task configuration and trajectory, provided that the information on the task geometry as well as appropriate linear transformations are given.

In real-world robot manipulation tasks, it may not be possible to obtain *exact geometric information* about a mechanism. Furthermore, the linear transformations for performing velocity propagations are usually difficult to formulate. In such cases, the method described in Sections 7.2 and 7.3 becomes handy to use.

7.5 Relative Motion Method of Analyzing Velocities

In the preceding sections, we have shown a qualitative geometric reasoning method formulated especially for the analysis of motion relationships in linkage mechanisms. Although the method may further be modified to handle more general CSI mechanisms, the resulting algorithm will conceivablely be complicated and dependent on the mechanisms being analyzed.

In this section, we will discuss a more general approach to deriving the qualitative description of linear velocities in CSI mechanisms based on individual bodies' *relative motions.* Typically, the information required is a description of the mechanism's configuration specifying qualitative positions (e.g., angular positions in the case of four-bar linkages) with respect to a set of *local reference frames* (i.e., relative coordinate systems as given in Definition 5).

From the definition of relative motion, we know that an absolute velocity may be expressed in terms of a sequence of velocities relative to an absolute velocity. In such a case, we say that the absolute velocity satisfies a *velocity constraint equation.* The fundamental idea of qualitative analysis with the relative velocity method is that, since we can find a set of motion vectors (see Figure 6.21) which qualitatively indicates the direction of relative motion, and an actual velocity vector is in fact proportional to its corresponding motion vector, we can write a velocity constraint equation describing the motion of a kinematic chain in terms of relative motion vectors. Furthermore, by evaluating and selecting sets of *vector modifiers* in the equation, we will be able to qualitatively determine both relative and absolute linear velocities.

7.5.1 Axioms and Theorems in Revolute or Prismatic-Pairing Body Motion

Of the various methods for transmitting motions, revolute and prismatic pairing methods are of the most interest in qualitative kinematics. Examples of mechanisms using these methods are linkages. In linkages, the motion of one link relative to another satisfies a certain constraint imposed by their intermediate pairs, and the velocity can be determined given the link's relative instantaneous position. In other words, it is possible to describe the constrained motion of a mechanism composed of such links in terms of the sum of individual links' relative motions.

Before we can qualitatively analyze the motions of a CSI mechanism using the relative motion approach, we will first formulate some fundamental axioms, theorems, and constructions concerning the motion of CSI mechanism components.

Revolute-Pairing Bodies

Suppose that body A is connected to body B by means of a revolute pair. In this case, the motion of A relative to B may be described in terms of the motion of A with respect to a reference frame on B originated at the rotational axis. In the foregoing discussion, we will use Cartesian coordinate systems as the relative reference frames. The relative instantaneous angular position of a given point on body A is defined as the smallest non-negative angle formed by the x-axis and the line segment passing through the point and its rotational axis. Hence, no matter in which quadrant the line segment lies, its relative angular position (θ) is always within the range of $[0, \frac{\pi}{2}]$.

Axiom 7.5.1 *Let a point on the body A be in rotation with respect to a reference frame and let l be the line segment passing through the point and its rotating axis. If the rotation is counterclockwise, then when l is in the first quadrant, the motion vectors corresponding to the set of qualitative angles can be described as in Table 7.1.*

Table 7.1. Relative motion vectors of a rotating point in quadrant I.

$\theta_I =$	va^-	va^+	a^-	a^+
(m_x, m_y)	(-s, vl)	(-m, l)	(-l, m)	(-vl, s)

Theorem 13 (Change of direction). *In Axiom 7.5.1, if the point rotates in an opposite direction, then the corresponding motion vectors will have the same magnitudes as before, but with opposite directions.*

Theorem 14 (Symmetrical property of a circular motion). *In Axiom 7.5.1, if l is in the second quadrant, then the direction of the y-coordinate components (m_y) in the corresponding motion vectors will change from positive to negative. If l is in the third quadrant, then the x and y components in the corresponding motion vectors will both change directions. If l is in the fourth quadrant, then the x components in the corresponding motion vectors will change direction.*

In general, it is always possible to determine the constrained motion of a mechanism from its reversed motion.

Theorem 15 (Inversion of a constrained motion). *Suppose that some constrained relative motion in a constrained closed-loop kinematic chain, A, is given. If one link of A moves over its entire range of motion, but with an opposite driving direction at each position, then the motions of all links in A reverse their directions.*

Prismatic-Pairing Bodies

The relative motion between two bodies A and B of a prismatic pair can be described in the same way as that of the revolute pair. A reference frame for the motion of A is fixed on B.

Axiom 7.5.2 *If the x-axis of the Cartesian system is parallel to a common tangent on the contact surface, then the motion of A relative to the frame can be described in terms of A's relative motion vector,* $(\pm vl, 0)$.

Axiom 7.5.3 *The prismatic motion of A relative to B is equivalent to the rotation of A relative to B with its center at infinity.*

Having understood the relative motion vectors of revolute-pairing and prismatic-pairing mechanism components, we can readily determine the constrained motion of an intermediately connected (e.g., linkage) mechanism, the details of which will be shown in Chapter 7.5.

7.5.2 Kinematic Modeling

In order to derive a velocity constraint equation, we will first find the relative motion vectors at the pairs of *links* (here the term link is used in a general sense). In general, corresponding to a specific chain progression in a derived mechanism graph, there exists a directed kinematic chain, $p_0 \to^{l_1} p_1 \to^{l_2} p_2 \to^{l_3} \ldots \to^{l_n} p_n$, where $p_{k-1} \to^{l_k} p_k$ denotes that link l_k is directed from lower pair p_{k-1} to p_k. In this chain, the local relative coordinate system for link l_k will be centered at the pairing contact p_{k-1} on l_{k-1}. In other words, the determination of positions for local reference coordinate systems in a mechanism will depend on the direction we choose for the chain progression. Given a set of relative reference frames, the derivations of relative motion vectors in relation to specific pairing contacts can be based on the axioms and theorems presented in Section 7.5.1.

As we know, an actual velocity vector has the same direction as its corresponding motion vector and their magnitudes are proportional to each other. Therefore, having obtained the motion vectors of a set of connected links, we can further find a constraint equation of the actual velocity vectors. In doing so, we may apply the following two theorems of constrained kinematic chains.

Theorem 16 (Loop postulate). *The algebraic sum of relative actual velocity vectors associated with the consecutive lower pairs of links in a simple closed-loop kinematic chain is zero.*

From Theorem 16, it is possible to further derive the following theorem:

Theorem 17 (Vertex postulate). *The actual velocity vectors of two links with respect to the same frame are equal at their lower-pairing contact.*

7.6 Qualitative Analysis of Relative Velocities

In this section, we discuss how to derive the qualitative description of motion of any specific link given the kinematic model of a CSI mechanism expressed in terms of an actual velocity constraint equation.

7.6.1 Solving Velocity Constraint Equations

The essence of qualitative reasoning about the motion of a CSI mechanism lies in the use of a heuristic search technique to modify the qualitative values of velocity vectors in a constraint equation initialized by motion vectors. The problem of heuristic search for appropriate velocity values can be stated as follows:

Given an initial representation of a velocity constraint equation, as expressed in terms of relative motion vectors, determine for each motion vector a sequence of modifiers such that the resulting vectors best satisfy the equation. This set of vectors is considered a qualitative solution of the velocity equation and therefore gives the absolute or relative velocities of links in the mechanism.

Here, the vectors that best satisfy the velocity equation are defined as those which, as compared to others resulting from *further* applying modifiers, yield the smallest error with respect to the equation. In order to obtain the overall best solution, the heuristic search is carried out in such a way that at each iteration all the possible modifiers are evaluated and those that can give a temporary best solution with respect to the previous vectors are selected. During each modification of velocity vectors, the derivations and evaluations of qualitative vectors are constructed from the inference rules of qualitative vector operation. The modified velocity vectors are termed *intermediate velocity vectors* or temporary velocity vectors.

7.6.2 An Algorithm for Determining Linear Velocities

An algorithm for determining linear velocities of a CSI mechanism utilizing the relative motion representation is given as follows:

Algorithm LINEAR_VELOCITY

Input: A representation of the mechanism's configuration in terms of the instantaneous position of each link with respect to some local reference frame at its lower pair.

Output: The desired velocity vector of a given link with respect to a fixed or moving link.

1. Derive a mechanism graph representation of the CSI mechanism.
2. Determine the independent loops in the graph which correspond to the constrained kinematic subchains in the mechanism.
3. Find the subchain which contains a link whose relative velocity at a certain pair is given.
4. Divide the subchain into two distinct chain progressions directed from the fixed link to the known pair.

5. For each chain progression, according to the given direction, express the velocity at the known pair in terms of the relative velocities of consecutive links. Connect these two expressions into a velocity constraint equation (Theorem 17).
6. For each chain progression, find the relative motion vectors of links utilizing the axioms and theorems presented in Section 7.5.1.
7. Transform the actual velocity vector terms in the original velocity constraint equation into corresponding modified motion vectors. If a relative velocity term is the given velocity then write its qualitative value.
8. Modify the set of motion (or intermediate) vectors in the new equation by using vector modifiers until the resulting vectors yield the smallest qualitative error in the original constraint equation. In each step of modification, all combinations of possible modifiers are evaluated and the (temporarily) best one is selected and applied.
9. Let the set of resulting intermediate vectors be the qualitative solution of the original velocity constraint equation. If the desired relative velocity between two links is within the current loop, then find its qualitative vector value by adding or subtracting the consecutive relative velocities in the given progression direction; else find the absolute velocity value of the link shared by another independent loop and consider the subchain corresponding to the new loop back to Step 4.

It should be noted that the *temporarily best modifiers* for a set of intermediate velocity vectors, as mentioned in Step 8, are defined such that

$$max(|E_{xi}|, |E_{yi}|) = min\{max(|E_{xj}|, |E_{yj}|)\}$$

where E_{xj} and E_{yj} denote the x and y velocity-component errors of the constraint equation, respectively, resulting from applying one of the four modifiers, j. E_{xi} and E_{yi} denote the errors resulting from applying temporarily best modifier i.

7.7 An Example

In this section, we present an example of qualitative reasoning about instantaneous linear velocities of a linkage mechanism with the relative motion method.

Example 7.3: Relative Motions in a Quick-Return Mechanism

The linkage to be analyzed is a quick-return mechanism, as shown in Figure 7.7, where the velocity of point d is desired and V_{b_2} is given

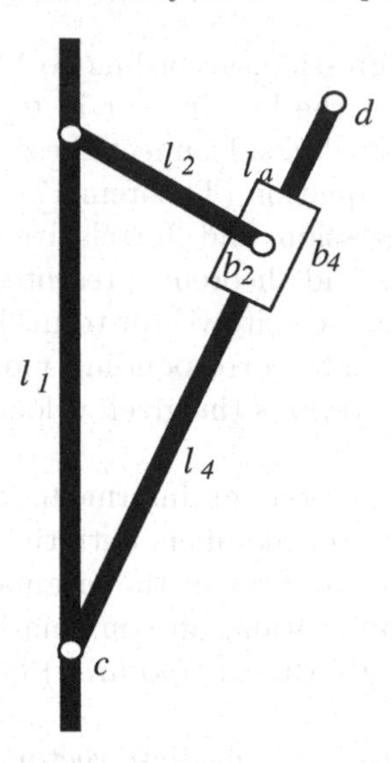

Fig. 7.7. A quick-return mechanism.

as a qualitative row vector (l, l). It can be noted that if the velocity at point b_4 which lies on the link l_4 is known, then V_d can be inferred by comparing the distances from the axis c to b_4 and to d. Therefore, the subgoal of the velocity analysis becomes the determination of linear velocity at b_4.

To begin the analysis, we represent the mechanism in an equivalent mechanism graph. Since component 3 is constrained to slide along link l_4, we can obtain an equivalent linkage mechanism by adding an imaginary link l_a between l_2 and l_4. The corresponding mechanism graph is given in Figure 7.8. It is obvious that the derived graph contains only one independent loop.

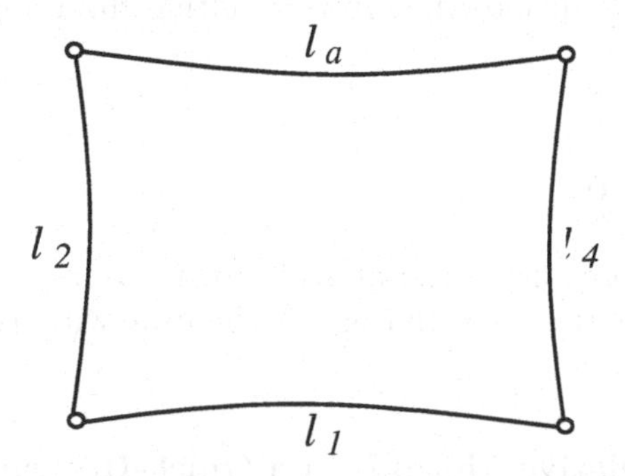

Fig. 7.8. The mechanism graph of a quick-return mechanism shown in Figure 7.7.

The next step is to construct a velocity constraint equation from the graph. We divide the loop into two distinct chains from one fixed joint to the known joint b_2 and, for each of the two chains, write V_{b_2} in terms of the sum of pairwise relative velocities. As the linear velocity at the endpoint of link l_2 relative to the fixed link l_1 is given, $V_{b_2 \leftarrow l_2}$, i.e., the velocity derived from the chain containing l_2, will be written in terms of the known qualitative row vector (l, l). Consequently, by Theorem 17 (vertex postulate), we can write a velocity constraint equation for this particular closed-loop mechanism as follows:

$$V_{b_2 \leftarrow l_a, l_4 \ldots} = V_{b_2 \leftarrow l_2} \tag{7.2}$$

or

$$V_{l_4/l_1} + V_{l_a/l_4} = (l, l) \tag{7.3}$$

where V_{l_4/l_1} and V_{l_a/l_4} denote the velocities of links l_4 and l_a relative to l_1 and l_4, respectively.

We know that the velocities V_{l_4/l_1} and V_{l_a/l_4} are proportional to their corresponding relative motion vectors m_{l_4/l_1} and m_{l_a/l_4}. Therefore, Eq. 7.3 can further be approximately rewritten as

$$\lambda_1 m_{l_4/l_1} + \lambda_2 m_{l_a/l_4} = (l, l) \tag{7.4}$$

where λ_1 and λ_2 denote the series of qualitative vector modifiers to be found. The relative motion vectors, corresponding to the given configuration, are shown in Figure 7.9. They are derived straightforwardly from Axioms 7.5.1, 7.5.2, and Theorem 13.

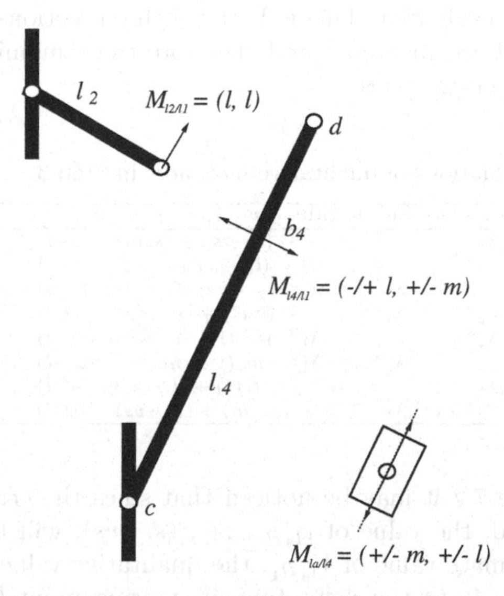

Fig. 7.9. The relative motion vectors of links in the mechanism of Figure 7.7.

Table 7.2. The modifications of motion (and intermediate velocity) vectors.

Steps	$\lambda_1 m_{l_4/l_1}$	$\lambda_2 m_{l_a/l_4}$	$V_{b2\leftarrow l_2}$	Errors
0	$(-l,m)$	(m,l)	(l,l)	$(-vl,m)$
1	$Inverse(-l,m)$	$Identity(m,l)$	(l,l)	$(m,-m)$
	$\Rightarrow (l,-m)$	$\Rightarrow (m,l)$		
2	$Decrease(l,-m)$	$Identity(m,l)$	(l,l)	$(s,-s)$
	$\Rightarrow (m,-s)$	$\Rightarrow (m,l)$		
3	$Decrease(m,-s)$	$Identity(m,l)$	(l,l)	$(vs,-vs)$
	$\Rightarrow (s,-vs)$	$\Rightarrow (m,l)$		

Having obtained Eq. 7.4, the next step of velocity analysis is to evaluate the possible combinations of the predefined modifiers and assign the best suitable set to the equation. The criterion is that the intermediate vectors resulting from applying modifiers should yield the smallest error in Eq. 7.3. This step is repeated until the error cannot be further reduced. Table 7.2 shows such an iterative process and Table 7.3 gives the details of the qualitative modifier evaluations in Step 3. In the tables, the qualitative inferences involved are based on the rules given in Section 7.5.1 and the error of Step i is defined as follows:

$$E_i = {\lambda_1}^i v_{l_4/l_1}^{(i-1)} + {\lambda_2}^i v_{l_a/l_4}^{(i-1)} - V_{b-2\leftarrow l_2} \tag{7.5}$$

where ${\lambda_1}^i$ and ${\lambda_2}^i$ denote the qualitative vector modifiers being applied in Step i, and $v_{l_4/l_1}^{(i-1)}$ and $v_{l_a/l_4}^{(i-1)}$ denote the intermediate vectors resulted from the iterative Step $i-1$. Note that in Step 3 the modifier *Inverse* is not evaluated. This is because the directions of velocities have been modified in Step 1 and therefore the remaining steps will deal only with magnitudes.

Table 7.3. The evaluations of qualitative modifiers in Step 3.

Steps	Increase	Decrease	Identity	$v_{l_4/l_1} + v_{l_a/l_4}$	$V_{b_2\leftarrow l_2}$	Errors
$3a$	${\lambda_1}^3$	${\lambda_2}^3$	-	$(l,-m)+(s,m)$	$(l,-l)$	$(s,-l)$
$3b$	${\lambda_1}^3$	-	${\lambda_2}^3$	$(l,-m)+(m,l)$	$(l,-l)$	$(m,-m)$
$3c$	-	${\lambda_1}^3$	${\lambda_2}^3$	$(s,-vs)+(m,l)$	$(l-,l)$	$*(vs,-vs)$
$3d$	${\lambda_2}^3$	${\lambda_1}^3$	-	$(l,vl)+(s,-vs)$	$(-l,l)$	$(s,0)$
$3e$	${\lambda_2}^3$	-	${\lambda_1}^3$	$(l,vl)+(m,-s)$	$(l-,l)$	$(m,-vs)$
$3f$	-	${\lambda_2}^3$	${\lambda_1}^3$	$(m,l)+(m,-s)$	$(l,-l)$	$(s,-s)$
$3g$	${\lambda_2}^3$,${\lambda_1}^3$	-		$(l,vl)+(l,-m)$	$(-l,l)$	$(l,-s)$
$3h$	-	${\lambda_2}^3$,${\lambda_1}^3$		$(s,m)+(s,-vs)$	$(-l,l)$	$(0,-s)$

From Table 7.2 it may be noticed that since the error cannot be further reduced, the value of v_{l_4/l_1}, i.e., $(s,-vs)$, will be considered as the approximate value of V_{l_4/l_1}, the qualitative value of the linear velocity of link l_4 at the instantaneous pairing point b_4 relative to fixed link l_1. Therefore, by comparing the distances from the fixed axis c to b_4 and to d, we can determine the qualitative value of the linear velocity at d. As given in the instantaneous configuration of this problem, distance cd is almost twice as long as distance cb_4, which is to say that the magnitude of V_d is similarly twice as large as that of V_{b_4}. Hence, we can derive the qualitative value of V_{b_4} to be $(l, -s)$. This step is readily understood by following the discussion presented in Chapter 6.

7.8 Summary

In this chapter, we have shown how to analyze the velocity relationships of a linkage mechanism given its dimensional specifications. We have developed an algorithm for the velocity analysis, which applies the kinematic concept of instantaneous axis. This approach, although computationally efficient, is to some extent limited to the analysis of linkage-like mechanisms.

Later in this chapter, we showed how to utilize the relative motion vector representation of mechanism components and to generate solutions by resolving qualitative motion constraint equations. Although this approach appears to be less computationally efficient than the qualitative trigonometric reasoning approach, it is more applicable in solving general CSI mechanism problems.

The algorithms described in this chapter are designed particularly for solving *instantaneous* velocity problems. It should be noted that these algorithms can also be extended to handle the kinematic state transitions of a moving CSI mechanism. In such a case, the analysis should be preceded by a step consisting of partitioning the value of an input displacement into a quantity space and computing the set of corresponding qualitative configurations, as mentioned in the previous chapters.

8

How to Plan Robot Motions

This chapter describes how the qualitative spatial planning technique (Section 5.3) and the simulated-annealing technique (Section 5.4) can be incorporated to solve the problem of configuration *planning* for *under-constrained planar mechanisms* (i.e., $dof > 1$).

In particular, two typical cases of spatial path planning for planar *robot-like* mechanisms are considered. The first one is concerned with planning a *low-cost collision-free* path for a single mobile polygonal object (e.g., mobile robots), whereas the second is a more general case of the path finding problem which deals with an articulated open-chain linkage mechanism (e.g., a robot manipulator) in a polygonal environment. Here, the notion of *low-cost* is defined in terms of the *length*, *passage clearance*, and *orientation cost* of the path.

The mobile object is a rigid body of convex polygonal shape, whose position and orientation relative to a reference frame can be specified by $\mathbf{C}=(x, y, \theta)$. The open-chain linkage is a kinematic chain made of n rigid links $\{L_1, L_2, ..., L_n\}$. Any two consecutive links are connected by either a revolute or a prismatic joint. In general, an n-link open-chain linkage requires n parameters to describe its instantaneous configuration, $\mathbf{C}=(\theta_1, \theta_2, ..., \theta_n)$. The configuration space of an under-constrained mechanism is the space spanned by its independent variables.

8.1 An Overview of the Method

The method of configuration planning for under-constrained planar mechanisms involves two major steps. The first step is to apply a heuristic search technique to find a global *route* that qualitatively satisfies certain optimality criteria. The route is represented as a sequence of disjoint m-closure regions. The second step is to use the simulated-annealing algorithm to perform local configuration search within the global route obtained.

The complete method can be summarized in the following algorithm:

Algorithm UC_CONFIG_PLAN

Input: A mobile under-constrained mechanism and a polygonal environment, with exact geometry (i.e., shapes, positions, and orientations), and initial and goal configurations of the mechanism being specified.
Output: A sequence of instantaneous configurations of the mechanism connecting the initial and the goal configurations.

1. **Free-space partition:** Partition free-space $\mathcal{E} - \mathcal{P}$ into a set of disjoint regions with m-edges.
2. **Qualitative route:** Find a qualitative route between the initial and the goal configurations, using the technique of qualitative spatial planning.
3. **Quantitative path segments:** Plan an exact path by applying `ANNEAL` to find the path segment from the initial configuration to the first m-edge configuration in the qualitative route and the path segment from the last m-edge configuration in the route to the goal configuration.
4. **Path composition:** Connect these two path segments by a sequence of precomputed m-edge path segments, corresponding to the segments in the qualitative route.

8.2 Qualitative Route Planning in the m-Edge Partitioned Euclidean Free-Space

Qualitative spatial planning is concerned with searching for a *low-cost route* for a moving mechanism (e.g., a planar object), where the cost is measured with respect to the route length, passage clearance, and orientation cost, as defined in Sections 5.3.1 and 5.3.2. Such a route provides a sequence of regions within which the *actual optimal paths* are likely to exist. In other words, the goal of the qualitative planning is to eliminate free-space regions where the actual paths are not feasible or will lead to infeasible regions or *dead-ends*, and to find the possible low-cost routes from which the actual paths can be found. In order to achieve this goal, the partition of the Euclidean free-space should meet the following three criteria:

1. Infeasible free-space regions can be identified.
2. Relative costs of a route can be checked from certain qualitative measurements over the representation.
3. The derivation of an exact path from the obtained qualitative route is not too costly.

This section shows how the m-closure region partitioning described in Chapter 4 satisfies the above requirements. The *object-environment* instance used throughout this section is defined as follows:

Let A denote a convex polygonal object whose configuration is specified by $\mathbf{C} = (x, y, \theta)$. A lies in a polygonal environment, P, composed of a set of n stationary non-intersecting convex polygonal objects $\{P_1, P_2, ..., P_i, ..., P_n\}$.

8.2.1 Eliminating Dead-End Regions

As has been proven in *classical geometry*, if two sets of planar points are convex, then their sum or difference is also convex [74]. In the above-mentioned *object-environment*, since convex polygons A and P_i are convex sets, the $\mathcal{C}$-space obstacle due to P_i is also a convex polygon [138]. This property of convex polygons will be used to prove the following theorem.

Theorem 18 (Clearance in $\mathcal{C}$-space). *The clearance in free $\mathcal{C}$-space of a convex polygonal object A, due to a polygonal environment P, is monotonically related to the clearance in P itself, where the clearance in P is measured by the minimum Euclidean distance between two polygonal objects. On the other hand, in the $\mathcal{C}$-space, the clearance is measured by the minimum Euclidean distance between two $\mathcal{C}$-space obstacle polygons with respect to a given orientation θ of A.*

Proof: Based on Lozano-Pérez's algorithm [138], it is possible to *grow* (compute) a $\mathcal{C}$-space obstacle for each of the stationary objects in P with respect to A. It is known that the grown $\mathcal{C}$-space obstacles must be convex. Let the minimum distance between two separate $\mathcal{C}$-space obstacles corresponding to P_i and P_j be $d_{\mathrm{obs_{ij}}}$ (in $\mathcal{C}$-space), and the minimum distance between P_i and P_j be $d_{\mathrm{P_{ij}}}$ (in P). The corresponding m-edge between P_i and P_j is denoted as $m_{\mathrm{P_{ij}}}$. If $d_{\mathrm{P_{ij}}}$ increases, i.e., P_j performs a translation along $m_{\mathrm{P_{ij}}}$ in the direction opposite to P_i, then by rigidity, all the points on the $\mathcal{C}$-space obstacle due to P_j will also perform a similar translation. Consequently, the distance between any two points on P_i and P_j respectively will increase. By definition, $d_{\mathrm{obs_{ij}}}$ must also increase. The above analysis is applicable to any given orientation θ. ∎

It immediately follows from Theorem 18 that if P_i and P_j are moving closer to each other, the two $\mathcal{C}$-space obstacles will contact each other, either before or exactly when the m-edge $m_{\mathrm{P_{ij}}}$ degenerates to a single point. At their contact, the two $\mathcal{C}$-space obstacles will form two concave vertices.

The minimum distance between the two $\mathcal{C}$-space obstacles will be called a *$\mathcal{C}$-space m-edge.* The contact or intersecting points between the $\mathcal{C}$-space obstacles can be regarded as the special case of a $\mathcal{C}$-space m-edge. In fact, they correspond to the dead-ends in the Euclidean free-space.

Observation 8.2.1 (Dead-ends in $\mathcal{C}$-space) *During the $\mathcal{C}$-space-based path search, if the local minima (i.e., dead-ends) are separated from the goal by a $\mathcal{C}$-space m-edge, then the shorter the $\mathcal{C}$-space m-edge, the narrower the escape passage and, consequently, the more costly to escape.*

Based on Observation 8.2.1 and Theorem 18, some insights into the Euclidean free-space can be gained.

Observation 8.2.2 (Hardness of dead-end escaping) *The cost of moving out of a dead-end region is created by the narrow passage in the Euclidean free-space.*

From Observations 8.2.1 and 8.2.2, it can be concluded that the m-edge partition of the Euclidean space will facilitate the identification and avoidance of dead-ends as well as the narrow passages leading to them.

8.2.2 Representing Path-Segment Invariants

Geometrically constrained motion of a convex polygonal object A in the Euclidean free-space $\mathcal{E} - \mathcal{P}$ (as opposed to the kinematically constrained motion discussed in Chapter 6) can be characterized if certain additional constraints are imposed.

The following two theorems are self-evident from the definition of *optimality* constraints.

Theorem 19 (Invariant optimal paths I). *If an optimization on the minimum traveling distance, the maximum clearance, and the minimum rotational maneuvers is achieved, the exact paths for A, connecting a given initial configuration* $\mathbf{C}_{\text{init}}$ *and a given goal configuration* $\mathbf{C}_{\text{goal}}$ *in a polygonal environment* P *(i.e., a static environment), are invariant, no matter how many times they are computed.*

Theorem 19 can be further generalized to the case where the initial and goal configurations are specified by finite sets.

Theorem 20 (Invariant optimal paths II). *Given* C_{init}, $C_{\text{goal}} \subseteq C_{\text{free}}$, *the exact optimal paths for A from* C_{init} *to* C_{goal} *(i.e., from* $\mathbf{C}^i_{\text{init}} \in C_{\text{init}}$ *to* $\mathbf{C}^j_{\text{goal}} \in C_{\text{goal}}$*) are invariant, no matter how many times they are computed.*

In a polygonal environment P, there are locations where *different paths* of a mobile object A are likely to pass through. These locations are called *critical locations.* If *all the critical locations* of the Euclidean free-space are known, an exact path between an initial location and a goal location can be readily composed by computing a path segment from the initial location to a critical location, a sequence of path segments between critical locations, and a path segment from the last critical location to the goal location.

From Theorems 19 and 20, it is known that the optimal path segments between the critical locations are invariant. Therefore, if the configurations at these locations are finite, the invariant path segments can be precomputed and stored before *planning* time. This will significantly reduce the cost of planning path segments (i.e., *searching* possible configurations).

While the exact critical locations may be hard to find, *approximate* places of such locations can be easily identified:

Observation 8.2.3 (Less freedom at narrow passages) *In a polygonal environment P, the configurations of a mobile polygonal object A at narrow passages are geometrically more constrained than those at broader passages (i.e., fewer significantly different configurations at narrow passages).*

As discussed in Section 5.3.2, the narrow passages can be characterized by m-edges. This implies that in planning an exact path of A from its qualitative route, if the route is represented as a sequence of connected m-closure regions, the *exact path* may be found by *connecting invariant path segments* between some configurations in the neighborhood of the m-edges.

8.2.3 An Algorithm for Finding Qualitative Routes

Given that $[LOC(\mathbf{C}_{\text{init}})] = m_{\text{init}}$ and $[LOC(\mathbf{C}_{\text{goal}})] = m_{\text{goal}}$ where m_{init} and m_{goal} are the m-closure regions in which the initial and goal configurations lie, respectively. A qualitative route from m_{init} to m_{goal} in an m-edge partitioned free-space can be represented as a sequence of connected m-closure regions that satisfies certain predefined *optimality criteria*. Generally speaking, either an initial or a goal configuration may span more than one m-closure region. In such cases, the region to be considered is the one for which a mobile object is heading.

As discussed in Chapter 5, an m-edge partitioned Euclidean free-space can be represented as a weighted connectivity graph, where each node denotes an m-closure region and each arc denotes a connection between two adjacent m-closure regions. Given such a graph, the problem of finding a short and wide route can be reduced to that of searching the graph in a depth-first fashion with the heuristic measurements to order the search steps as nodes are expanded. Such a search is known as *hill-climbing.* Here, the choice of applying the hill-climbing search is mainly due to the consideration that the qualitative cost functions are well suited for *stepwise cost evaluations* as in hill-climbing. Other strategies [191] may involve the summation of the stepwise costs, which cannot be unambiguously computed without introducing some arbitrary numerical scales (see Appendix C).

The hill-climbing search algorithm can be formally stated as follows:

Algorithm QUALITATIVE_ROUTE

1. Construct a queue that initially contains only the start region m_s.
2. Select the next region m_i from the queue that has the shortest distance to the goal [*length*]. If ties occur, resolve in favor of a greater *passage clearance* (or less *orientation cost*).
   a) If the queue is empty, terminate the algorithm, and announce failure.

   b) If m_i is a goal region, terminate the algorithm, and announce success.
   c) Otherwise, remove m_i from the queue and calculate [*length*] for each connected region of m_i; add these regions to the front of the queue, sorted according to [*length*].
3. Go to Step 2.

The *passage clearance* and *orientation cost* can be computed according to their definitions given in Section 5.3.2.

8.3 Constructing Exact Paths from Qualitative Routes

In planning an exact path for a mobile object A from its qualitative route, if the route is represented as a sequence of connected m-closure regions, the *exact quantitative path* may be found by *connecting invariant path segments* between some configurations in the neighborhood of the m-edges (see Section 8.2.2). This section presents a *formal representation* as well as a *search technique* for the path segments.

8.3.1 The Composition of an Exact Path

The formal representation of an exact path for A in an m-edge partitioned Euclidean free-space is given as follows:

A path of A, denoted as $\mathcal{P}_{\mathrm{A}}$, from $\mathbf{C}_{\mathrm{init}}$ to $\mathbf{C}_{\mathrm{goal}}$ is a continuous map $p : [k-1, k] \rightarrow C, k = 1, 2, ..., i, ..., n$, with $p(0) = \mathbf{C}_{\mathrm{init}}$, $p(i) = \mathbf{C}_{\mathrm{m}}$, and $p(n) = \mathbf{C}_{\mathrm{goal}}$, where $\mathbf{C}_{\mathrm{init}}$ and $\mathbf{C}_{\mathrm{goal}}$ are the *initial* and *goal* configurations of A, respectively, and $\mathbf{C}_{\mathrm{m}}$ is the configuration of A at an m-edge. $\mathcal{P}_{\mathrm{A}}$ is therefore composed of three types of path segments, namely,

1. A starting path segment from $\mathbf{C}_{\mathrm{init}}$ to an m-edge configuration $\mathbf{C}_{\mathrm{m}}$:
$$\mathcal{P}_{\mathrm{A_{start}}} = \{p(x) \mid x \in [0, 1]\} \tag{8.1}$$
2. A sequence of m-edge *connected* path segments from one m-edge configuration to another:
$$\mathcal{P}_{\mathrm{A_{m-edge}}} = \{p(x) \mid x \in [1, n-1]\} \tag{8.2}$$
3. An ending path segment from an m-edge configuration $\mathbf{C}_{\mathrm{m}}$ to $\mathbf{C}_{\mathrm{goal}}$:
$$\mathcal{P}_{\mathrm{A_{end}}} = \{p(x) \mid x \in [n-1, n]\} \tag{8.3}$$

Definition 20 (m-edge configurations). *An m-edge configuration of a mobile object A is defined as the configuration of A where the interior region of A intersects an m-edge, and the joint angle formed by the m-edge and the shortest diameter (see Figure 8.1a) of A is* `Acute` *(*`Acute` *is defined in Definition 11).*

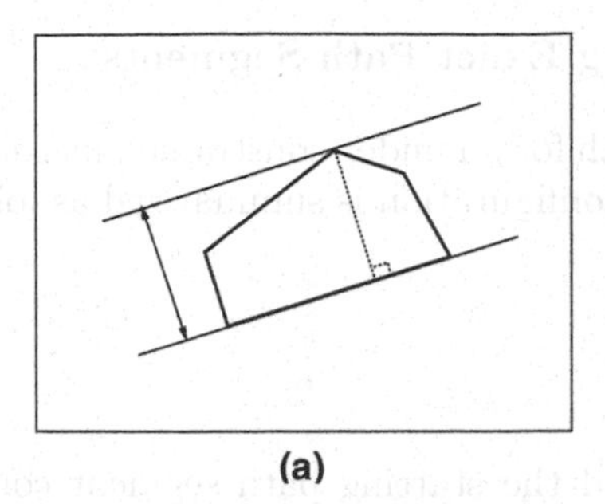

(a)

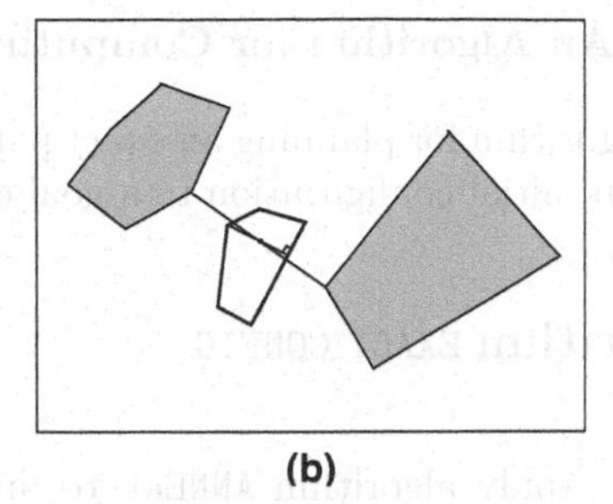

(b)

Fig. 8.1. (a) The shortest diameter of a convex polygonal object. (b) A small change to the configuration of the polygonal object where its shortest diameter is aligned with an m-edge.

It is obvious from Definition 20 that the m-edge configuration of A with respect to a certain m-edge will not be unique. In the current work, such a configuration is generated in two steps, as illustrated in Figure 8.1b: first, aligning the shortest diameter of A with the m-edge (of two possible directions) and their mid-points and, second, making a *small* random change (i.e., motion) to the current configuration of A, $\mathbf{C}_i$, such that the joint angle of the m-edge and the diameter of A is `Acute`.

8.3.2 Randomized Search for Exact Path Segments

Let m_i and m_j be two m-edges which bound the same m-closure region, and $\mathbf{C}_i$ and $\mathbf{C}_j$ be the corresponding m-edge configurations at m_i and m_j, respectively. These two m-edge configurations are said to be *adjacent*, since their m-edges bound the same m-closure region.

For each pair of adjacent m-edge configurations, $\mathbf{C}_i$ and $\mathbf{C}_j$, an exact connecting path (i.e., a chain of configurations between $\mathbf{C}_i$ and $\mathbf{C}_j$) can be created by applying the `ANNEAL` algorithm as presented in Section 5.4. During the *simulated annealing*, each configuration $\mathbf{C}_k$ of an n-degree-of-freedom mechanism is specified by a set of n configuration variables, and is generated *within the m-closure region bounded by* $\mathbf{C}_i$ *and* $\mathbf{C}_j$. The objective function E is defined as the Euclidean distance between configurations $\mathbf{C}_k$ and $\mathbf{C}_{\text{goal}}$, where $\mathbf{C}_{\text{goal}}$ is the goal configuration. In the above case, if a path from $\mathbf{C}_i$ to $\mathbf{C}_j$ is desired, then $\mathbf{C}_{\text{goal}} = \mathbf{C}_j$.

During the local configuration search within an m-closure region, the control temperature function $T(i)$ for the annealing process (see Eq. 5.10 in `ANNEAL`) determines how a *local minimum configuration* (i.e., a dead-end) will be escaped; that is, if $T(i)$ is non-zero, then there exists a certain probability that a configuration which is further from the goal than the previous configuration may also be accepted.

8.3.3 An Algorithm for Computing Exact Path Segments

The algorithm for planning an exact path for an under-constrained mechanism from an initial configuration to a goal configuration is summarized as follows:

Algorithm `EXACT_CONFIG`

1. Apply algorithm `ANNEAL` to find the starting path segment connecting $\mathbf{C}_{\text{init}}$ and the *first* m-edge configuration in the qualitative route.
2. Retrieve all the offline computed m-edge *connected* path segments corresponding to a sequence of connected qualitative route segments.
3. Apply algorithm `ANNEAL` to find the ending path segment connecting the *last* m-edge configuration in the qualitative route to the goal configuration $\mathbf{C}_{\text{goal}}$.

With a graphic representation of an m-edge partioned free-space, a qualitative route is represented by a partial ordered list of *nodes* (i.e., m-closure regions). *Each arc* connecting two nodes specifies the adjacency relationship between two regions which is *associated with a particular m-edge that divides the two regions.* Hence, an m-edge configuration in a qualitative route (e.g., the *first* m-edge configuration) can be located.

8.4 Graphical Simulations

A path planner has been implemented to validate the above algorithms[1]. The planner can deal with two basic types of robot-like under-constrained mechanisms, namely a *mobile object with two translational and one rotational degrees of freedom* and a *highly-redundant manipulator with several rotational degrees of freedom.*

The input given to the planner consists of the specifications of a set of polygonal obstacles and an under-constrained mechanism with its initial and goal configurations. The planner generates, as an output, a sequence of configurations, called a path, which connects the initial and goal configurations. The derived path should *qualitatively* satisfy the following optimality criteria:

1. the route of the path is the shortest among all the possible ones and

[1] The qualitative route planning is carried out by a module implemented in NASA/CLIPS, whereas the local quantitative path search is written in C language. The graphical user interface is developed using OSF/Motif X Windows subroutines.

2. the route of the path maximizes the passage clearance and minimizes the orientation cost among those motions satisfying (1).

A number of experiments with various robot environments has been conducted. During path planning, it is assumed that the motion of a robot is not limited by any dynamic constraints. Furthermore, the general path-planning problem is specialized by assuming that the obstacles and the mobile object are of *convex* polygonal shape.

8.4.1 Examples

Example 8.1: Path Planning for a Single Mobile Object

In the example shown in Figure 8.2, the unfilled convex polygons represent the obstacles in the environment, and the two filled convex polygons represent the initial and the final configurations of the same mobile object. An exact path connecting these two configurations is desired. It is assumed that *the exact shapes of the objects, as well as the exact geometry (positions and orientations) of the polygonal environment are given.*

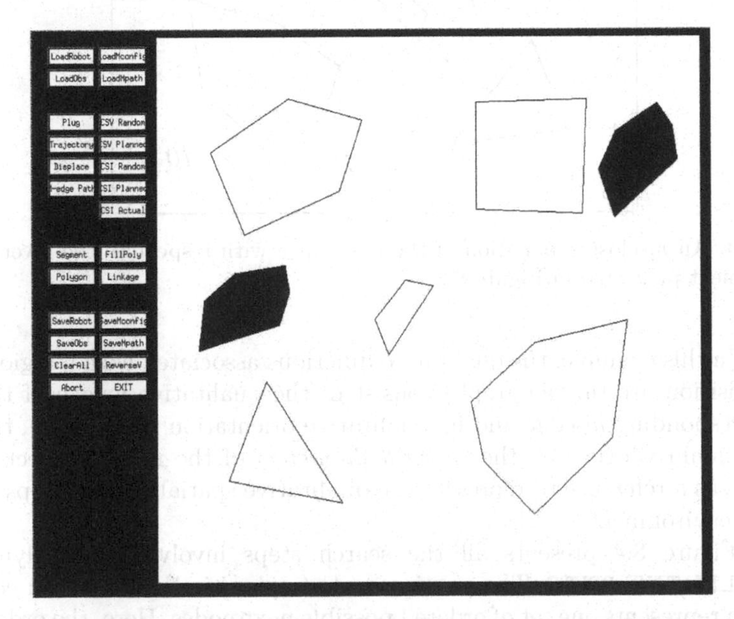

Fig. 8.2. A mobile planar object in a polygonal environment. The filled polygon on the left displays the *initial* configuration of the mobile object, and the one on the right displays the *goal* configuration.

Qualitative Route Planning

According to the UC_CONFIG_PLAN algorithm, the first step (**Free-space partition**) of path planning is to construct an m-edge partition of the free-space. The procedure for such a construction is based on the proof of Lemma 1. The resulting partition, as given in Figure 8.3, is represented as a graph.

This graphic representation of the free space is used as the basis to perform qualitative route search, using the QUALITATIVE_ROUTE algorithm. As mentioned in Section 5.3.2, three cost functions can be used to measure the optimality of a qualitative route during the heuristic search, namely route length, passage clearance, and orientation cost.

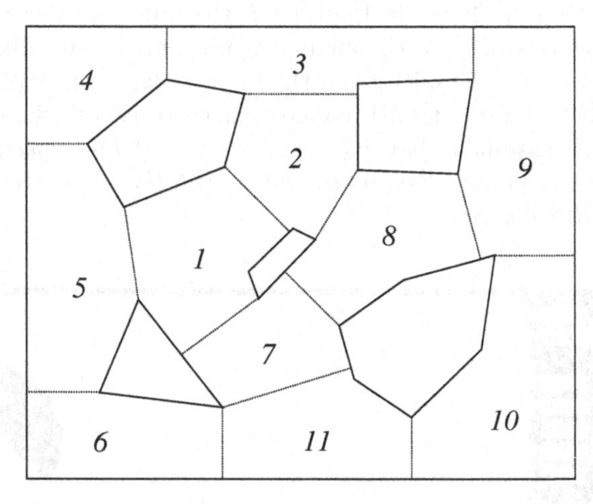

Fig. 8.3. An m-closure partition of the free-space with respect to the given robot environment presented in Figure 8.2.

In this example, the qualitative functions associated with a region-transition arc (in the graph) consist of the qualitative length of the corresponding m-edge and its qualitative orientation angle. Here, the minimum width (i.e., the *shortest diameter*) of the mobile object is used as a reference to represent the qualitative spatial relationships of the environment.

Figure 8.4 presents all the search steps involved in applying QUALITATIVE_ROUTE. The starting node (region) is Node 1. Each column represents one set of ordered possible next nodes. Here, the ordering of choices is determined by the results of the following *sequential evaluation* of the cost functions:

1. *the remaining distance to the goal region* (i.e., the minimum number of arcs to the goal from the next region),
2. *the passage clearance of the next arc*, and

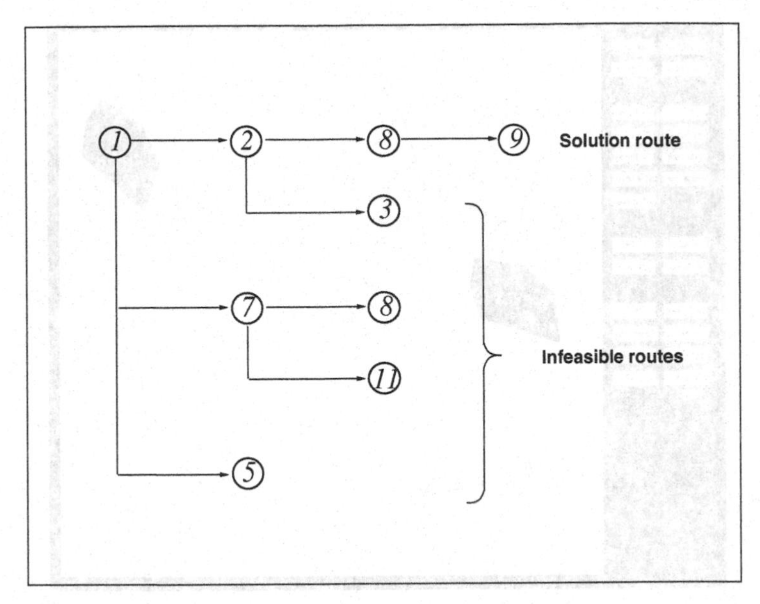

Fig. 8.4. Stepping through the qualitative route search with the `QUALITATIVE_ROUTE` algorithm.

3. *the orientation cost of the next arc.*

The resulting qualitative route is shown as a sequence of regions at the top of Figure 8.4. This route is depicted in Figure 8.5 as a sequence of dashed line segments.

Exact Path Derivation

Based on the result of qualitative spatial planning, i.e., a qualitative route, as shown in Figure 8.5, exact path derivation can be performed using the `EXACT_CONFIG` algorithm of Section 8.3.

The region in which the mobile object initially lies is considered to be 1 (note that region 5 can also be assumed since the initial configuration overlaps with both regions 1 and 5). The derived qualitative route specifies the *region transition* to be:

$$1 \longrightarrow 2 \longrightarrow 8 \longrightarrow 9$$

where region 9 is the region in which the goal configuration lies. Accordingly, the m-edge configurations can be identified in the following order, as shown in Figure 8.6:

$$\mathbf{C}_{m_{12}} \longrightarrow \mathbf{C}_{m_{28}} \longrightarrow \mathbf{C}_{m_{89}}$$

where $\mathbf{C}_{m_{ij}}$ denotes an m-edge configuration at the m-edge shared by both region i and region j.

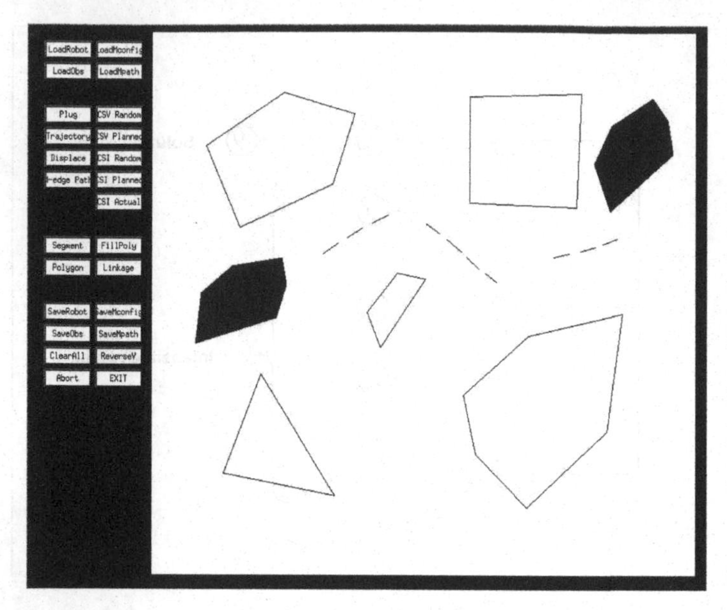

Fig. 8.5. A possible route (in dashed lines) found by using qualitative *hill-climbing* search.

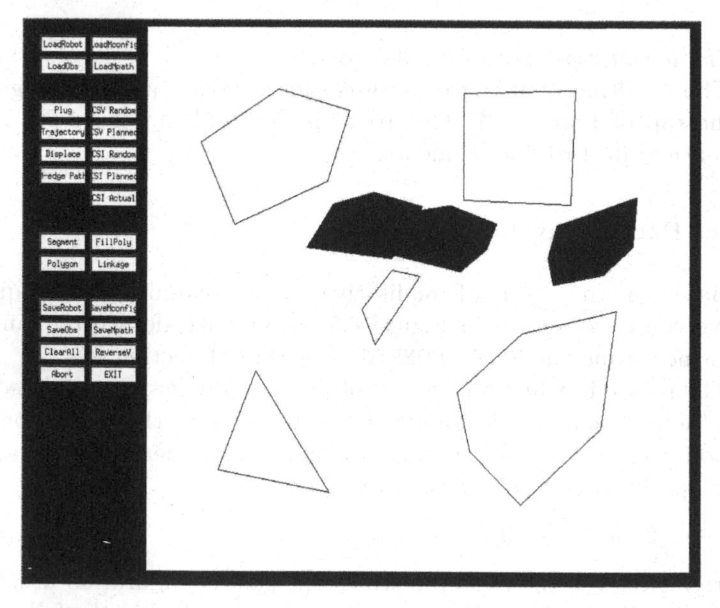

Fig. 8.6. A set of m-edge configurations identified from the qualitative route, as shown in Figure 8.5.

Having identified the above ordering of the m-edge configurations, the planner proceeds with the following path construction (as detailed in the EXACT_CONFIG algorithm):

1. Apply the ANNEAL algorithm to construct a path segment from $\mathbf{C}_{init}$ to $\mathbf{C}_{m_{12}}$.
2. Retrieve a sequence of two path segments, one connecting $\mathbf{C}_{m_{12}}$ to $\mathbf{C}_{m_{28}}$ and the other connecting $\mathbf{C}_{m_{28}}$ to $\mathbf{C}_{m_{89}}$.
3. Search for a path segment from $\mathbf{C}_{m_{89}}$ to $\mathbf{C}_{goal}$ using the ANNEAL algorithm.

Figure 8.7 shows the results of the first step. Figures 8.8 and 8.9 present the two respective path segments retrieved from a library of precomputed m-edge connected path segments. Figure 8.10 shows the ending path segment. The entire path composed of the above segments is presented in Figure 8.11.

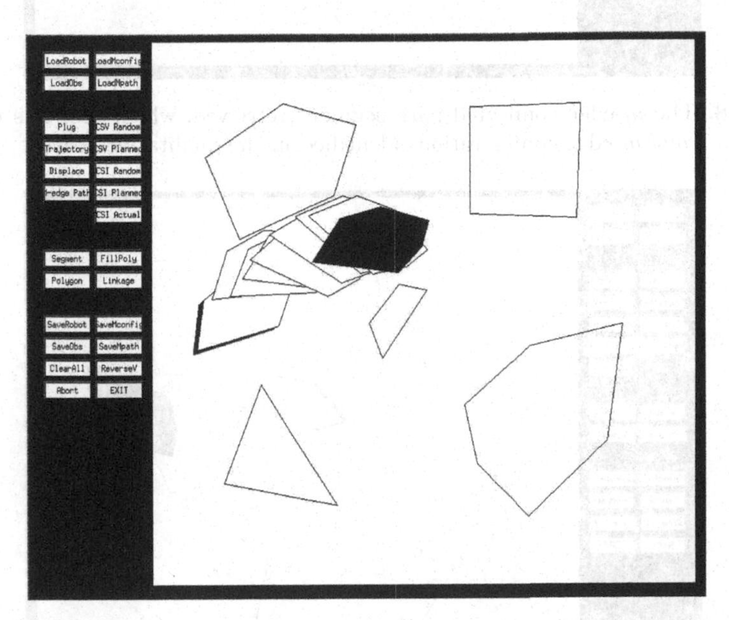

Fig. 8.7. The m-edge connected path segment from the *initial* configuration of the robot to its *first* m-edge configuration.

Example 8.2: Path Planning for a Highly Redundant Articulated Manipulator

This example is concerned with the construction of an exact path for a serially coupled manipulator which is composed of *six* links. Each link L_i is coupled to L_{i-1} and L_{i+1} only, L_0 being the (fixed) base

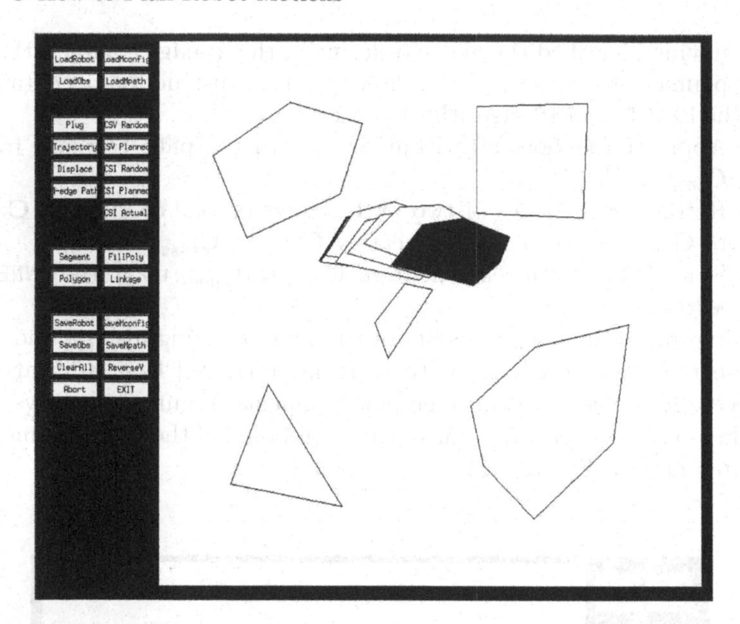

Fig. 8.8. The m-edge connected path segment retrieved, which connects the *first* and the *second* m-edge configurations identified in the qualitative route.

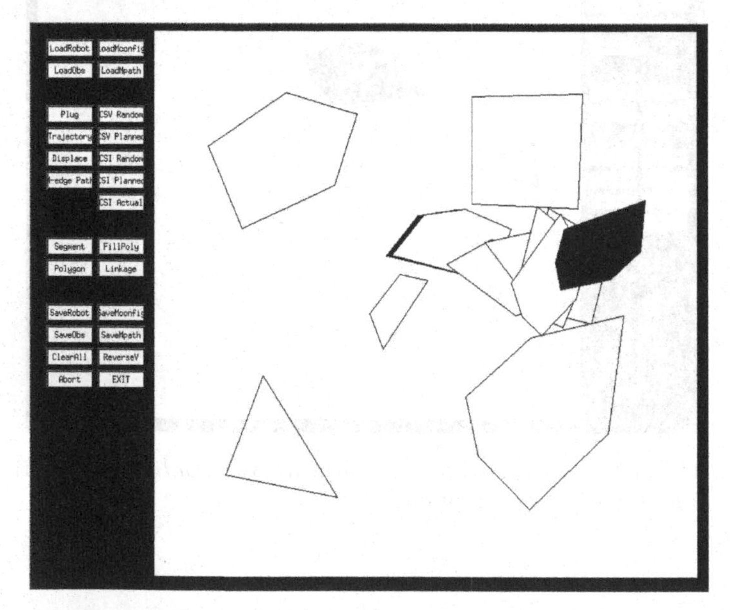

Fig. 8.9. The m-edge connected path segment retrieved, which connects the *second* and the *third* m-edge configurations identified in the qualitative route.

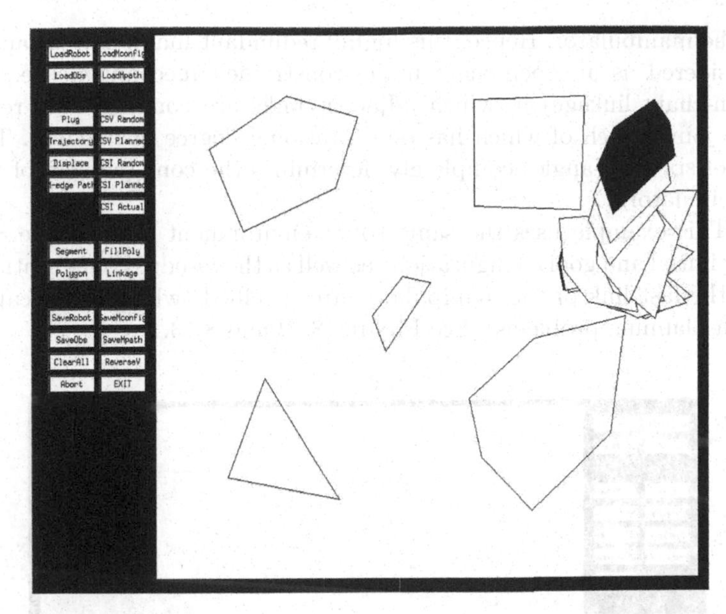

Fig. 8.10. The m-edge connected path segment for *exiting* an m-edge configuration.

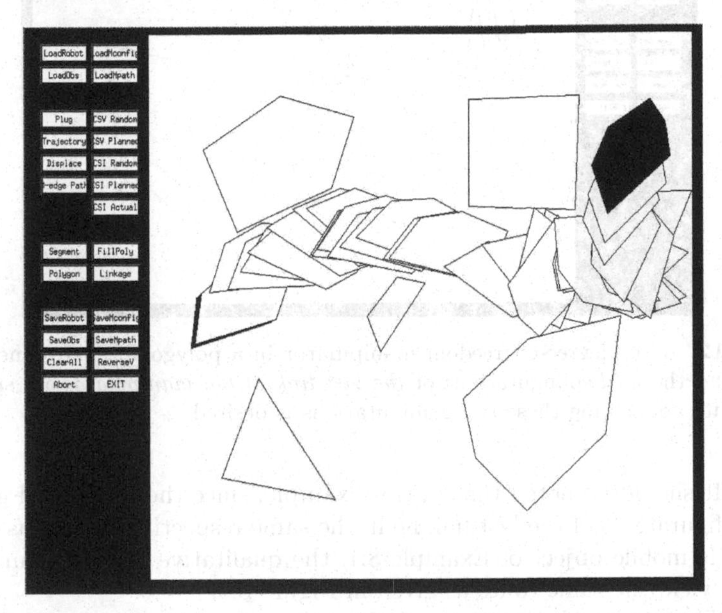

Fig. 8.11. The complete path from $\mathbf{C}_{init}$ to $\mathbf{C}_{goal}$ found by using algorithm `UC_CONFIG_PLAN`.

of the manipulator. Hence, this highly redundant manipulator can be considered as an open-chain under-constrained mechanism (i.e., an open-chain linkage) in which adjacent links are connected by revolute joints, each of which has one rotational degree of freedom. The set of six joint angles completely determines the configuration of the manipulator.

This example uses the same robot environment as Example 8.1. The initial and goal configurations as well as the m-edge configurations for the last link of the manipulator are specified (which is typical in path-planning problems). See Figures 8.12 and 8.13.

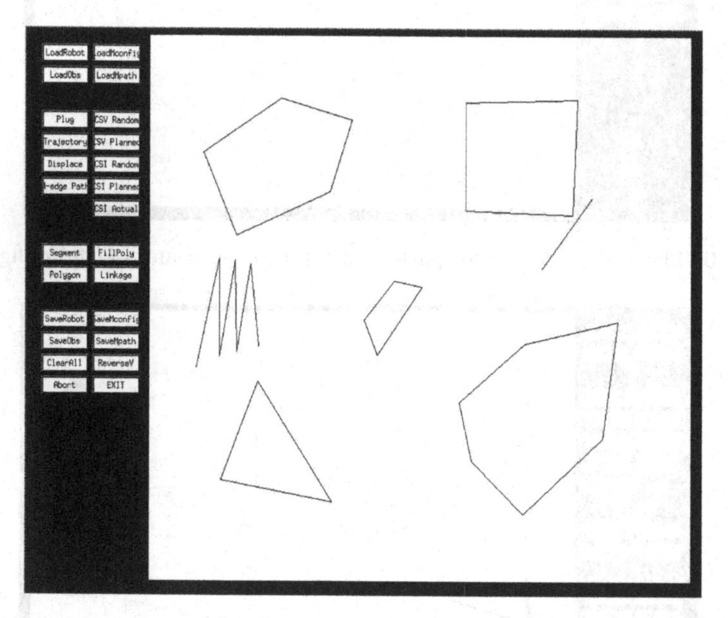

Fig. 8.12. A six-degree-of-freedom manipulator in a polygonal environment. The *initial* and the *goal* configurations of *the last link of the manipulator* are specified, and a path connecting these two configurations is desired.

It should be noted that in this example, since the initial and goal configurations of the last link lie in the same respective regions as the single mobile object of Example 8.1, the qualitative spatial planning will yield the same route as given in Figure 8.5.

The results of *quantitative path* construction are presented in Figures 8.14 through 8.18, which show the path segment from the initial configuration to the first m-edge configuration, the first m-edge connected path segment, the second m-edge connected path segment, the final path segment from the last m-edge configuration to the goal configuration, and the complete path found.

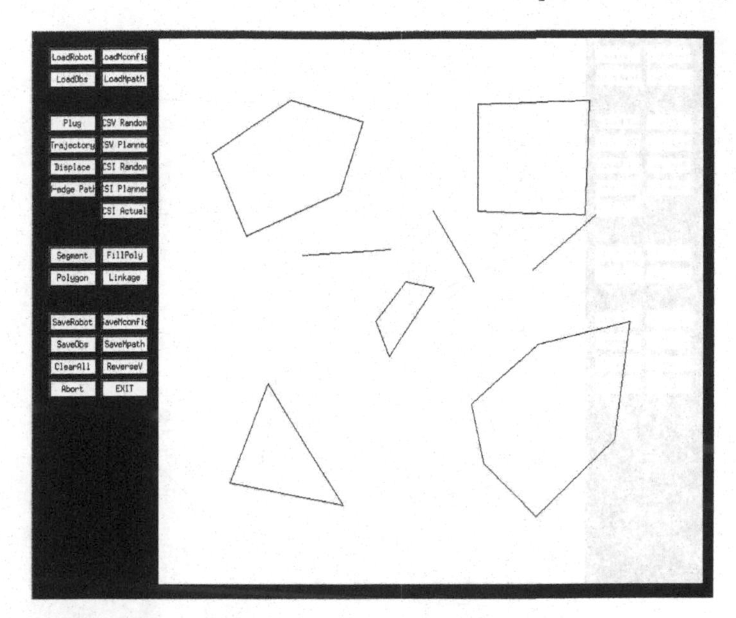

Fig. 8.13. The m-edge configurations of the manipulator of Figure 8.12, identified with respect to the qualitative route, as shown in Figure 8.5.

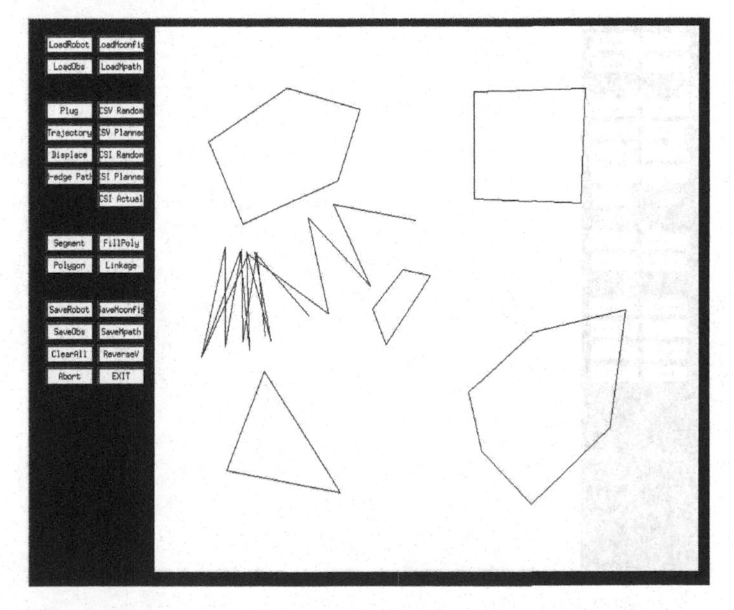

Fig. 8.14. The path segment of the manipulator of Figure 8.12, which connects the *initial* configuration of the manipulator to its *first* m-edge configuration.

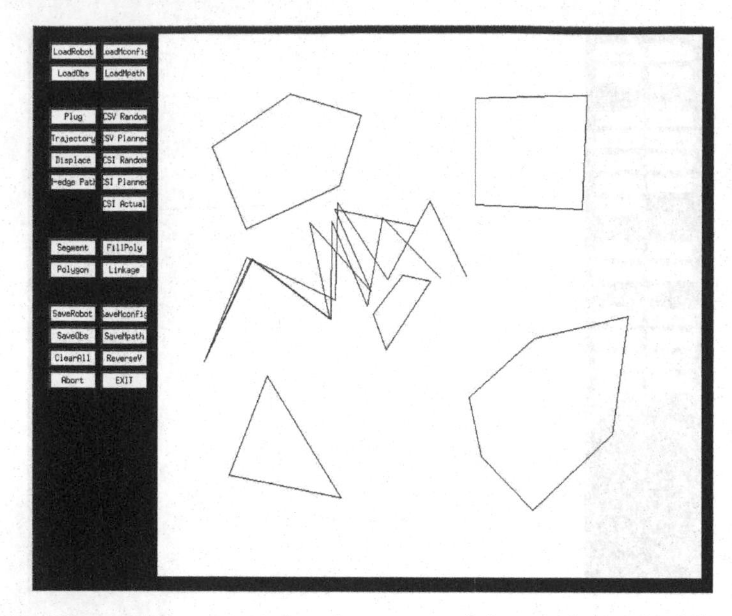

Fig. 8.15. The m-edge connected path segment of the manipulator of Figure 8.12, from the *first* m-edge configuration to the *second*.

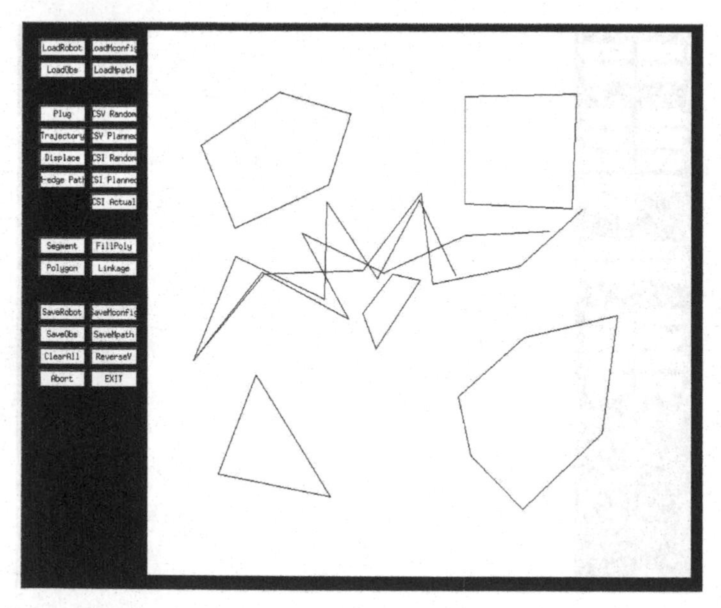

Fig. 8.16. The m-edge connected path segment of the manipulator of Figure 8.12, from the *second* m-edge configuration to the *third*.

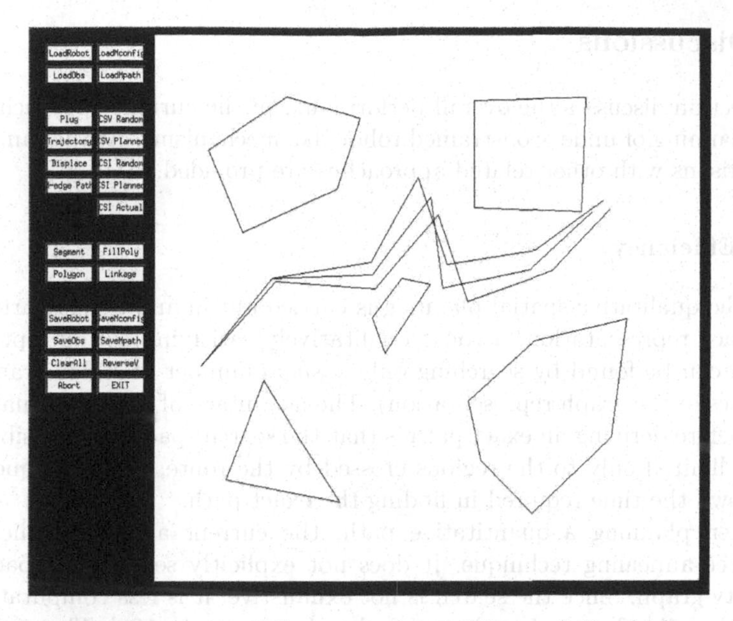

Fig. 8.17. The path segment of the manipulator of Figure 8.12, which connects the *last m-edge configuration* to its *goal* configuration.

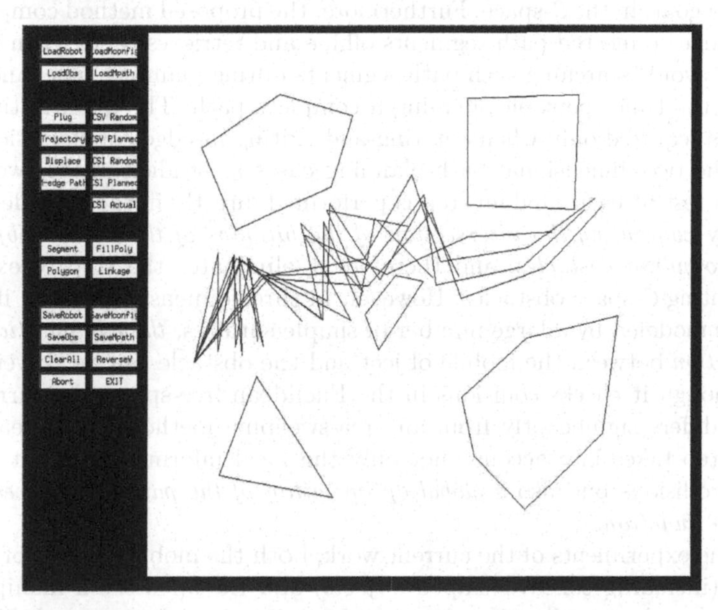

Fig. 8.18. The complete path of the manipulator of Figure 8.12, found with the `UC_CONFIG_PLAN` algorithm.

8.5 Discussions

This section discusses the overall performance of the current approach to the path planning of under-constrained robot-like mechanisms. At the same time, comparisons with other related approaches are provided.

8.5.1 Efficiency

Since the qualitative spatial planning is carried out in an m-edge partitioned free-space representation, a route qualitatively satisfying certain optimality criteria can be found by searching only a *small* number of region transitions (i.e., arcs in the graph representation). The advantage of finding a qualitative route before deriving an exact path is that the search space for a feasible path can be limited only to the regions crossed by the route, and consequently it cuts down the time required in finding the exact path.

When planning a quantitative path, the current approach relies on a simulated-annealing technique. It does not explicitly search a $\mathcal{C}$-space connectivity graph. Since the search is not exhaustive, it is less computationally costly than the $\mathcal{C}$-space-based spatial planning methods [138]. The simulated-annealing technique may be considered as a biased randomized search (biased toward a goal) in the $\mathcal{C}$-space. Furthermore, the proposed method computes all the m-edge connected path segments offline and retrieves them when needed. Thus, it avoids searching such path segments during planning time, and hence reduces the time spent on planning a complete path. The quantitative path search is required only when entering and exiting m-edge configurations.

In the two-dimensional path-planning cases as studied in this work, the collision test at each random step is performed directly in the Euclidean free-space by *computing the intersection of the bitmaps of the mobile object and its surrounding obstacles*, and therefore it eliminates the step of explicitly representing $\mathcal{C}$-space obstacles. However, in three-dimensional cases, if an obstacle is modeled by a large number of simple surfaces, *the computation of an intersection* between the mobile object and the obstacle can be slower.

Although it checks collisions in the Euclidean free-space, the current approach differs significantly from an area-sweeping method in that each proposed step takes into account not only the local information about the potential collisions but also *a global optimization of the path with respect to an objective function.*

In the experiments of the current work, both the mobile objects of various shapes (including concave shaped objects) and the open-chain manipulator-like linkage of more than five degrees of freedom have been tested. The simulation results have shown that the proposed qualitative-quantitative path search method is efficient.

8.5.2 Near-Obstacle Paths

It can be noted that the paths generated using the UC_CONFIG_PLAN algorithm may sometimes follow the boundaries of obstacles. Such paths can have the following advantages:

1. **Robustness:** If there exist certain geometric, sensory, or modeling uncertainties during the motion execution, the boundary-following paths will indicate the contact information for compliant motions.
2. **Locality:** In tightly constrained environments, contact configurations in the paths provide local references and bracing supports for the robot movements.

The drawback of the boundary-following paths is that if the robot task requires much *clearance* between the robot and the obstacles, such paths will not guarantee to provide it. Nevertheless, this drawback can be overcome by enclosing an obstacle inside an envelope and treating the envelope as one obstacle.

8.5.3 Comparison with Other Free-Space-Based Approaches

Brooks [18] has proposed a method of representing two-dimensional robot free-space as natural "freeways" between obstacles and an algorithm to find collision-free paths. The basic idea of the representation is to compute *candidate generalized cones*. Experiments with the freeway method have demonstrated that the algorithm works well in uncluttered environments, with less computation time than the general configuration-space method [18]. However, the drawback of this approach is that paths are restricted to follow the spines of the generalized cones chosen to represent the free-space.

Unlike the "freeway" approach, OĎuńlaing et al. [159] and Yap [193] have developed a method of using a *generalized Voronoi diagram* for motion planning. The Voronoi diagram of a set of obstacles in the robot environment is constructed by computing the locus of points that are equidistant from at least two of the obstacle boundaries. This approach of mapping the free-space onto a one-dimensional network of curves within that space is also known as *retraction*.

The current approach to the free-space partition utilizes the notion of m-closure regions. The m-closure representation has the following advantages over both the "freeways" partition and the generalized Voronoi diagram representation:

1. **m-closure region:** The m-closure region is characterized by a set of minimum-spanning edges, which are easy to construct.
2. **Qualitative route:** Qualitative route planning can be efficiently performed, given an m-closure representation.
3. **Region-based local search:** Exact path planning within a sequence of m-closure regions does not constrain a path to follow any one-dimensional network of exact lines or curves.

8.5.4 Comparison with Other Monte-Carlo Path-Planning Approaches

Barraquand and Latombe [7] have presented an approach of path planning based on randomized search within a *grid potential field* representation. The implemented planner can solve high-dimensional configuration-space problems. Given an initial configuration, it applies a *best-first* motion (a standard *informed search* technique) to follow the steepest descent of the potential, and terminates once a local minimum is reached. Next, the planner starts to execute a series of *random motions* in an attempt to escape from the local minimum. These steps are repeated until a goal configuration is reached. As indicated in [127], this approach involves computing the grid potential field which works fast only in low-dimensional space. The size of the grid increases exponentially with the dimension m of the robot's configuration space.

Furthermore, since the potential field does not tend toward infinity when the robot gets closer to an obstacle, the best-first motion does not guarantee to be collision-free. The planner must check collision for each progression from one configuration to another.

Glavina [72] has proposed an approach which is quite similar to the one by Barraquand and Latombe [7]. In his approach, *subgoal configurations* of a robot are randomly generated from the robot free-space, whenever local minima are found during the goal-directed *straight-and-slide* search in $\mathcal{C}$-space.

The path-planning method described in this chapter incorporates both qualitative *global route* planning and quantitative *local path* planning, and has the following advantages over the above two approaches: if *quantitative* spatial information about the robot and its environment is *not* available, the qualitative spatial planning step may still succeed in generating possible routes in terms of sequences of connected m-closure regions. In addition, the qualitative spatial planning step will effectively eliminate certain dead-end regions prior to the exact path planning.

In the proposed approach, the simulated-annealing-based local path search does not rely on configuration-space potential fields, nor configuration-space subgoals. It is carried out in a direct m-closure free-space representation, and is localized in the m-closure regions specified by a qualitative route. Furthermore, the exact path segments, such as the m-edge connected path segments, can be precomputed and (re)used in composing complete paths with different initial and goal configurations.

8.5.5 Limitations

The proposed method has the same limitation as the constrained mechanism analysis discussed in Section 6.4.2, namely the inexactness of the path and the *empirical* nature of the *control temperature for annealing* (i.e., the $T(i)$ function in Eq. 5.10 of `ANNEAL`). Also, it is necessary to ensure that valid starting and ending m-edge configurations are found before generating an m-edge connected path segment. Another drawback of path derivation based on simulated annealing is that there can be a large spectrum of spatial frequencies in the path. Smoothing of such a path may be required.

Since the path derivation is performed based on the *qualitative route* information, if the heuristic *search* fails to generate a route when in fact one exists, no exact path will be composed.

9 How to Make Spatial Measurements and Maps

Making perceptual hypotheses about the world using sensing measurements is a classical problem in robotics. Consider, for example, a mobile robot equipped with a ring of sonar (ultrasound) sensors. Given a particular *measurement*, i.e., a particular set of sonar values, we may variously hypothesize that an object is just to the right or just to the left of the robot. Depending on the hypothesis, the robot may then move straight ahead or turn left or right.

Practically, it is unlikely that a measurement will exactly match one of the possible perceptual hypotheses, and this suggests that a matching strategy based on Fuzzy Set (FS) theory may be appropriate, since we will be able to formalize the notion that a given measurement is *partial* evidence for each hypothesis. Such an approach is proposed in [35] for the wall-following behavior of a mobile robot. Distance estimates associated with the sonar data are classified in a fuzzy way as being 'very-near', 'near', 'somewhat-near', 'somewhat-far', 'far', or 'very-far', and then distinct control rules are defined for each case. However, no implementation is suggested.

In this chapter, we will define the mapping from measurements to perceptual hypotheses in terms of fuzzy relations. In principle, the mapping from measurement to perceptual hypothesis is straightforward as long as we trust each sonar value equally, given a suitable definition of the membership functions. But it is well known that the accuracy of the sonar sensor is the highest when the incident angle of the beam is normal to the surface of the object.

In different kinds of situations, sonar sensors may behave more or less reliably, i.e. sensing values may be more or less uncertain, and such uncertainty should be taken into account when deciding how to 'react' to the current perceptual hypothesis. For the case of a mobile robot, this may mean 'trusting' the hypothesis "no obstacle ahead" more or less in different kinds of situations. If there is much uncertainty associated with the hypothesis, then the robot may decide to advance or turn slightly in position and take another measurement instead of moving forward.

This will suggest the need for a new mapping to relate the uncertainty associated with the sonar values to the uncertainty associated with the perceptual hypothesis based on those values.

9.1 Mapping

In this section, we present a more precise statement of our problem. For the sake of argument, consider a robot equipped with N sensors, and let x_i^j be the value from sensor i at time instant j. Then, $\mathcal{X}^j = (x_1^j \ldots x_N^j)$ defines a measurement at time j. We will call this a point in "m-space". From $\mathcal{X}^j$, we now wish to obtain $\mathcal{H}^j$ corresponding to the perceptual hypothesis at time j. We will define the mapping as $\mathcal{H}^j = \phi(\mathcal{X}^j)$. Furthermore, we associate with each sensor value x_i^j an uncertainty $u(x_i^j)$. In this way, we may define the uncertainty of the measurement $\mathcal{X}^j$ as $\mathcal{U}(\mathcal{X}^j) = (u(x_1^j) \ldots u(x_N^j))$, denoted by $\mathcal{U}^{\mathrm{X}}$. We will call this "m-uncertainty". Finally, we define the uncertainty $\mathcal{U}(H^j)$ associated with the perceptual hypothesis $\mathcal{H}^j$ by the mapping $\mathcal{U}(\mathcal{H}^j) = \psi(\mathcal{X}^j, \mathcal{U}(\mathcal{X}^j))$ (or, in simplified notations, $\mathcal{U}^H = \psi(\mathcal{X}, \mathcal{U}^{\mathrm{X}})$).

The various mappings are summarized in the following commutative diagram. The semantics of the operators $\circ\mathcal{R}(x,h)$ and $\odot\mathcal{R}(x,h)$ will be defined shortly.

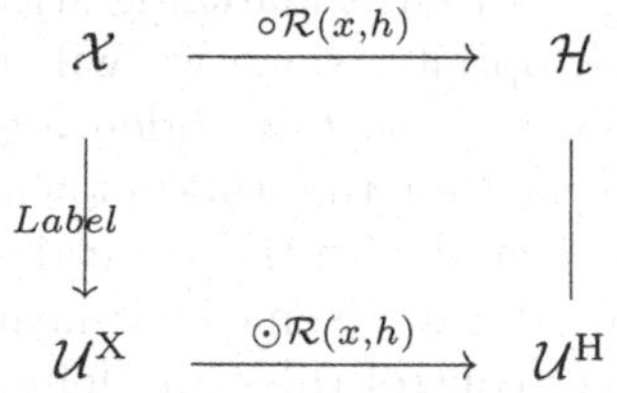

$$\begin{array}{ccc} \mathcal{X} & \xrightarrow{\circ\mathcal{R}(x,h)} & \mathcal{H} \\ \Big\downarrow {\scriptstyle Label} & & \Big| \\ \mathcal{U}^{\mathrm{X}} & \xrightarrow{\odot\mathcal{R}(x,h)} & \mathcal{U}^{\mathrm{H}} \end{array}$$

In the next section, we concentrate on the mappings ϕ and ψ and suggest how they may be defined in the context of FS theory.

For the sake of illustration, we will use a simple example throughout our discussion. We consider a robot equipped with $N = 3$ sonar sensors with the first pointing to the left, the second pointing to the front, and the third pointing to the right. We introduce the simple but effective heuristic proposed in [148] for associating uncertainty (or, inversely, accuracy) with a normalized sensor value x (distance, as determined by time of flight). The labeling scheme is defined as follows: *short* $\Rightarrow$ *LO* uncertainty, *medium* $\Rightarrow$ *MED*, and *long* $\Rightarrow$ *HI*, where the ranges $\{short, medium, long\}$ are specified by three non-overlapping intervals between 0.0 and 1.0.

In this example, we assume that there are two possible perceptual hypotheses of particular interest:

- h_1 : "obstacle on the left" and
- h_2 : "obstacle on the right".

9.2 m-Uncertainty and FS Theory

Fuzzy Set (FS) theory offers a systematic way of representing partial set memberships and handling inexact pattern matching [114]. More importantly, the compositional rule of inference enables us to make fuzzy implications and thereby to derive possible perceptual hypotheses. Related work on sensory data interpretation using fuzzy sets can be found in [59]. In fuzzy sets, the degree of membership in a fuzzy set is measured by a generalization of the characteristic function called membership function $g_{\mathcal{X}}(x) : X \rightarrow [0, 1]$ (g for grade of membership), where X is the universe of all possible members, i.e., the set of all possible vectors of sensor values. In the context of our current work, each $\mathcal{X}$ in m-space is considered as a fuzzy set, where the sensor value indicates the grade of membership for that specific sensor. Similarly, each $\mathcal{H}^j$ is represented as an array of possibilities – a fuzzy set defined in the universe of $H = \{h_1, h_2, \ldots, h_k\}$, i.e., the set of all possible perceptual hypotheses.

Within this framework, we can now construct the mapping from $\mathcal{X}$ to $\mathcal{H}$, utilizing the rules of fuzzy inferences. Let us assume that we have a list of fuzzy implication rules (referring to the three-sensor example):

- if $\mathcal{X} = (short, medium, long)$, then $\mathcal{H} = (1, 0)$ i.e., obstacle on the left and
- if $\mathcal{X} = (long, medium, short)$, then $\mathcal{H} = (0, 1)$ i.e., obstacle on the right.

Using new definitions *short* = 0.25, *medium* = 0.4, and *long* = 0.75, we can re-write the above rules as follows: (0.25,0.4,0.75) $\rightarrow h_1$ and (0.75,0.4,0.25) $\rightarrow h_2$. Then, given a particular measurement, say $\mathcal{X}^j = (0.2, 0.3, 0.6)$, we wish to find a new $\mathcal{H}^j$ as a result of fuzzy implications.

According to the Fuzzy Logic formalism, we implement each $\mathcal{X} \rightarrow \mathcal{H}$ via a fuzzy relation $\mathcal{R}$. Similar to the fuzzy set definition, the fuzzy relation is a fuzzy set defined in the Cartesian product universe. It is an extension of the crisp relation to include the membership grade and can be expressed as follows:

$$\mathcal{R} = \{g_{\mathcal{R}}(x, h)/(x, h) \mid (x, h) \subseteq X \times H\} \tag{9.1}$$

The membership grade can be derived through the use of the fuzzy set theoretic operator [196]:

$$g_{\mathcal{R}}(x, h)/(x, h) = \min\{1, 1 - g_{\mathcal{X}}(x) + g_{\mathcal{H}}(h)\} \tag{9.2}$$

Given a new $\mathcal{X}$, we apply the compositional rule of inference (i.e., generalized *modus ponens*) to assert that $\mathcal{H}$ in H induced by $\mathcal{X}$:

$$\mathcal{H} = \mathcal{X} \circ \mathcal{R} \tag{9.3}$$

where $\circ$ is the composition operator (i.e., max – min operator), defined as follows:

$$\max_{\mathbf{x}} \ \min\{g_{\mathcal{X}}(x), g_{\mathcal{R}}(x, h)\} \tag{9.4}$$

Now let us return to our example. Following the above formulation, we can readily derive the relation for the first fuzzy implication rule as below:

$$\mathcal{R}_1 = \begin{matrix} & h_1 & h_2 \\ x_1 \\ x_2 \\ x_3 \end{matrix} \begin{pmatrix} 1 & 0.75 \\ 1 & 0.6 \\ 1 & 0.25 \end{pmatrix}$$

Furthermore, we have

$$\mathcal{H}_1 = (0.2, 0.3, 0.6) \circ \begin{pmatrix} 1 & 0.75 \\ 1 & 0.6 \\ 1 & 0.25 \end{pmatrix} = (0.6, 0.3)$$

Similarly, with the second implication rule, we have

$$\mathcal{H}_2 = (0.2, 0.3, 0.6) \circ \begin{pmatrix} 0.25 & 1 \\ 0.6 & 1 \\ 0.75 & 1 \end{pmatrix} = (0.6, 0.6)$$

Finally, we combine the two results by using the fuzzy-intersection operator [196], which yields:

$$\mathcal{H} = \mathcal{H}_1 \cap \mathcal{H}_2 = (0.6, 0.3).$$

In other words, the measurement $(0.2, 0.3, 0.6)$ supports the perceptual hypothesis h_1 with degree 0.6 and h_2 with degree 0.3. In the next section, we describe a way of associating m-uncertainty with perceptual hypotheses $\mathcal{H}$.

9.3 Incorporating m-Uncertainty

Thus far, we have developed a representation of how the sensor values $\{x_i^j\}$ which define the current measurement, $\mathcal{X}^j$, give relative support to the possible perceptual hypotheses about the world, i.e. the mapping $\mathcal{H}^j = \phi(\mathcal{X}^j)$. In this section, we look more closely at how to take into account the time-varying uncertainty associated with the measurements the robot obtains about the world, in the context of FS theory developed in the previous section.

In this work, we assume that the uncertainty associated with x_i^j is propagated to the hypotheses during the max – min operation. More specifically, we propose a new compositional rule of m-uncertainty propagation, that is,

$$\mathcal{U}_i^{\mathrm{H}} = \mathcal{U}^{\mathrm{X}} \odot \mathcal{R} = \sum_{\mathrm{j}} \Delta_{ij}^{-1} \mathcal{U}_j^{\mathrm{X}} \tag{9.5}$$

where Δ_{ij} is the absolute difference between $g_{\mathcal{R}}(x_j, h_i)$ and $g_{\mathcal{X}}(x_j)$. $\odot$ is the proposed compositional operator.

In our three-sensor example, this rule is best illustrated as follows:

$$\mathcal{U}^{\mathrm{X}} = (u(x_1), u(x_2), u(x_3)) = (LO, LO, HI),$$

$$\mathcal{U}_1^{\mathrm{H}} = \mathcal{U}^X \odot \mathcal{R}_1 = \Big((1.68 * LO + 2.5 * HI),\ \ (5.15 * LO + 2.86 * HI)\Big),$$

$$\mathcal{U}_2^{\mathrm{H}} = \mathcal{U}^X \odot \mathcal{R}_2 \ = \ \Big((23.33 * LO + 6.6 * HI),\ \ (2.68 * LO + 2.5 * HI)\Big).$$

Finally, we use the fuzzy-union operator to compute $\mathcal{U}^H$ as follows:

$$\mathcal{U}^H = \mathcal{U}_1^{\mathrm{H}} \cup \mathcal{U}_2^{\mathrm{H}} = \Big((23.33 * LO + 6.6 * HI),\ (5.15 * LO + 2.86 * HI)\Big).$$

Or, with normalization,

$$\mathcal{U}^H = \Big((0.78 * LO + 0.22 * HI),\ \ (0.64 * LO + 0.36 * HI)\Big).$$

In other words, we associate with the measurement $(0.2, 0.3, 0.6)$ the m-uncertainty (LO, LO, HI). The resulting fuzzy sets $(0.78 * LO + 0.22 * HI)$ and $(0.64 * LO + 0.36 * HI)$ then define the association of m-uncertainty with h_1 and h_2, respectively.

9.4 Collective Spatial Map Construction

In the following sections, we will consider the problem of collective spatial modeling with a group of autonomous agents. The major challenges addressed are (1) how to enable the distributed agents to dynamically acquire their goal-directed cooperative behaviors in performing a certain task and (2) how to apply the methodology of group learning to handle ill-defined problems, e.g., problems without complete representations [137]. In response to those challenges, we have developed a novel evolutionary self-organization approach to collective spatial modeling, and demonstrated the implemented approach in tackling a specific problem, namely how to collectively build an artificial potential-field map of an unknown environment. In the chapter, we will present both the detailed formulation of our approach and the results of empirical validation in the context of spatial modeling with a group of agents.

In real-life applications, autonomous agents are often used to perform specific tasks in a distributed fashion [133]. The agents may be physically embodied, such as systems that efficiently manipulate some objects in a Cartesian environment from one designated location to another. The motivation behind our present work on distributed autonomous agent groups comes from the fact that the distributed multi-agent approach has a number of advantages over the traditional single complex systems approach, in that the groups can readily exhibit the characteristics of structural flexibility, reliability through redundancy, simple hardware, adaptability, reconfigurability, and maintainability [137].

The distributed agents locally interact with their environments in the course of collective spatial modeling. Responding to different local constraints received from their task environments, the agents may select and exhibit different behavioral patterns. The behavioral patterns may be carefully pre-defined or dynamically acquired by the agents based on certain learning mechanisms [133]. In the context of this chapter, we will focus on the latter case, where the agents are required to learn their behaviors for the purpose of collectively accomplishing a given task.

Using inexpensive off-the-shelf hardware, it is now possible to build mobile agents that can perform a variety of interesting tasks, such as wall-following and obstacle-avoidance. In this chapter, we describe an approach to evolving agent group behaviors in performing a cooperative spatial modeling task. The proposed approach begins with the modeling of local interactions between the agents and their environment, and then applies a genetic algorithm as a global optimization method for selecting the reactive motion behaviors of the individual agents with an attempt to maximize the overall effectiveness of collectively building a global spatial model.

The practical significance of this work lies in that the resulting behavior learning mechanisms for the autonomous agents can readily lead to real-life robotic applications, such as:

1. inspection of contaminated (e.g., radioactive) sites or facilities,
2. multi-agent navigation and searching in a hazardous area during a rescue mission,
3. security surveillance in a museum,
4. bomb disposal and toxic waste disposal, and
5. path planning in a container/cargo terminal or warehouse.

9.4.1 Related Work on Spatial Map Construction

Spatial modeling in an unknown environment is a challenging problem. Some of the earlier studies have tackled this problem by developing exact search algorithms that acquire graph-like representations of an unknown environment. An example of this approach will be the work by Betke et al. [11] on the piecemeal learning of an agent environment containing only convex objects. Others have addressed the problem by representing and identifying an unknown environment in terms of a set of basic geometrical primitives, such as line segments, circles, regions, landmarks, or local maps [10, 186]. The acquired maps are sometimes associated with uncertainty estimations in order to take into account the inherent uncertainty in sensors [161]. In such an approach, incremental learning algorithms such as Kohonen neural networks and Kalman filters are frequently applied [86, 95, 112, 118]. While the majority of the map-building studies deal with the problem of modeling two-dimensional unknown environments, some researchers have investigated the use of a self-organizing approach to reconstructing an unknown three-dimensional surface [5].

In our present work, we focus on the problem of dynamically constructing an artificial-potential-field representation of an unknown environment [105], without the explicit modeling of geometrical primitives. This will reduce the computational complexity involved in identifying the primitives, and at the same time provide a free-space representation that can be readily used for both global and real-time reactive motion planning [92]. Also, unlike the above-mentioned studies in which only a single mobile agent or sensing system is involved, our work addresses the issue of collective potential-map construction using a group of autonomous agents. Furthermore, we are interested in how to enable the distributed agents to develop cooperative behaviors in their collective spatial modeling. While some earlier studies have shown the navigation of agents in an unknown environment by using a set of pre-defined behaviors [73], our work takes one step further by offering a general evolutionary self-organization approach to the problems of group-oriented behavior learning and task performance that requires no explicit engineering of agent reactive behaviors.

9.4.2 The Problem

The problem of spatial planning was traditionally treated as an optimization problem, in which the configuration of an agent is represented in a parameter space, and a solution to this problem is computed by searching the parameter space in an attempt to satisfy a predefined cost function, such as the distance between the agent and a goal point. A significant limitation of this approach is that it is computationally too costly to generate new plans when dealing with dynamic environments that involve unexpected objects. As a more practical approach to real-time planning of collision-free motions for agents, the notion of artificial potential field (APF) was proposed by Khatib [103, 105]. The APF approach incorporates dynamic sensing feedback into agent control, and hence overcomes the aforementioned limitation by extending the reactiveness of the low-level motion control.

Artificial-potential-field (APF) theory states that for any goal-directed agent in an environment that contains stationary or dynamically moving objects, an APF can be formulated and computed, taking into account an attractive pole at the goal position of the agent and repulsive surfaces of the objects in the environment. This potential field can be expressed as follows:

$$\mathcal{P}_{apf}(x) = \mathcal{P}_{goal}(x) + \mathcal{P}_{obs}(x) \tag{9.6}$$

where $\mathcal{P}_{apf}(x)$, $\mathcal{P}_{goal}(x)$, and $\mathcal{P}_{obs}(x)$ denote the artificial potential field, the attractive potential from the goal, and the repulsive potential from the objects, respectively; x denotes a set of independent parameters, called operational coordinates, that describe the position and orientation of the agent.

A possible expression of attractive potential will be

$$\mathcal{P}_{goal}(x) = -\frac{1}{2}k_p(x - x_{goal})^2 \tag{9.7}$$

where k_p is a positive gain.

An example of repulsive potential is given as follows:

$$\mathcal{P}_{obs}(x) = \begin{cases} \frac{1}{2}\eta\left(\frac{1}{\rho} - \frac{1}{\rho_0}\right)^2 & \text{if } \rho \leq \rho_0 \\ 0 & \text{if } \rho > \rho_0 \end{cases} \tag{9.8}$$

where ρ_0 is a distance threshold, beyond which no repulsive force will be received by the agent.

Generally speaking, $\mathcal{P}_{obs}$ is chosen such that $\mathcal{P}_{apf}$ is a nonnegative continuous and differentiable function that tends to infinity when x approaches the surface of an object and tends to zero when x approaches the goal position, x_{goal}.

Given Eq. 9.6, the force, resulting from the APF at x, can therefore be derived:

$$F_{apf} = -\nabla[\mathcal{P}_{apf}(x)] \tag{9.9}$$

where ∇ denotes a gradient.

The above expression tells us that applying artificial potential field $\mathcal{P}_{apf}(x)$ to an agent can here be realized by using F_{apf} as a command vector to control the agent in operation space (as the motion of the agent can be decoupled in operation space [104]). Under such a control, the agent will be able to avoid objects as the repulsive force in the potential field "pushes" it away into the valleys of the field, and at the same time be able to move toward a goal position as the attractive force in the potential field "pulls" it in the direction of a global zero-potential pole.

The question that remains is how the APF methodology can be used if the agent environment concerned is not given as *a priori* knowledge. In such a situation, it will be essential to dynamically derive a numerical potential-field representation based on the sensory data obtained during the interaction between the agent and its environment.

The problem that we consider in this work can be stated as follows: How can the task of building a potential-field representation of an unknown agent environment be carried out by a group of autonomous agents? The environment that the agents are in is a bounded two-dimensional Cartesian space that contains a number of objects. The artificial-potential-field map to be built should reflect the spatial clearance in the environment. Our goal is to enable the distributed agents to cooperatively perform the map-building task, based on few sensory measurements. That is to dynamically construct the potential-field map as efficiently as possible.

The distributed agents for our potential-field-map-building task utilize local sensors to get proximity information about their surrounding objects or other agents. They can move, within the free-space of the environment, a certain step in a given direction.

9.5 Self-Organization of a Potential Map

9.5.1 Coordinate Systems for an Agent

In our implementation, a two-dimensional Cartesian coordinate frame (x, y) is constructed for a given agent environment. As a spatial model of this environment, a global potential-field map that is to be built by a group of agents is overlaid on top of this frame.

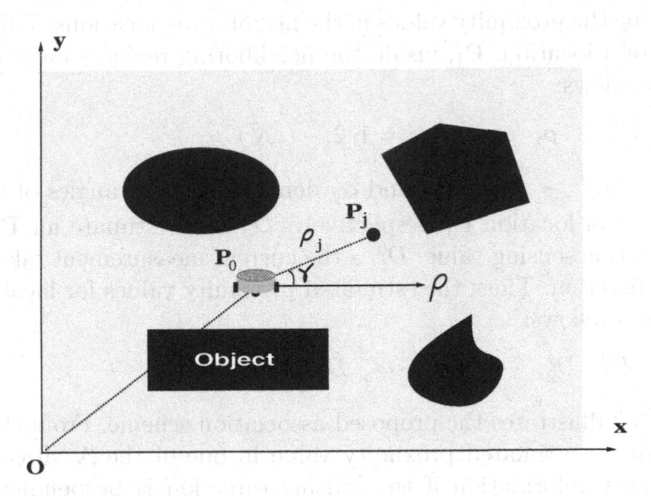

Fig. 9.1. The coordinate systems of an agent in an environment.

The position of each agent for map-building in the environment is described using Cartesian coordinates $\mathbf{P_O}(x_O, y_O)$. With respect to the current position of the agent, a relative polar coordinate frame is also constructed. Figure 9.1 shows the relationship between the Cartesian coordinates of the agent in the environment and its relative polar coordinates. From the figure, it is clear that a new location, $\mathbf{P_j}(x_j, y_j)$, can be expressed with respect to $\mathbf{P_O}(x_O, y_O)$ as follows:

$$\begin{bmatrix} x_j \\ y_j \end{bmatrix} = \begin{bmatrix} x_O \\ y_O \end{bmatrix} + \rho_j \begin{bmatrix} cos\alpha \\ sin\alpha \end{bmatrix} \tag{9.10}$$

where α and ρ_j denote the relative polar angle and polar radius of location $\mathbf{P_j}$, respectively.

9.5.2 Proximity Measurements

An agent measures its distances to the surrounding objects of its environment from several directions. In measuring the proximity information, the agent takes a set of $\mathcal{N}$ measurements with a resolution of $\frac{2\pi}{\mathcal{N}}$ per reading. These measurements are recorded in a sensing vector, $\mathcal{S}_0$, with respect to location $\mathbf{P_0}$, that is,

$$\mathcal{S}_0 = \left[D_1^0,\ D_2^0,\ \cdots,\ D_i^0,\ \cdots,\ D_{\mathcal{N}}^0\right]. \tag{9.11}$$

9.5.3 Distance Association within a Neighboring Region

Having represented the proximity information at $\mathbf{P_0}$ with vector $\mathcal{S}_0$, the agent then associates this information to its adjacent locations in the environment by estimating the proximity values in the neighboring locations. The estimated proximity of a location, $\mathbf{P_j}$, inside the neighboring region to a sensed object is given as follows:

$$\hat{D}_i^j = D_i^0 - \rho_j \cdot cos\beta \quad (i = 1, 2, \cdots, \mathcal{N}) \tag{9.12}$$

where $\beta = \alpha_0^{(i)} - \alpha_j$. $\alpha_0^{(i)}$ and α_j denote the polar angles of the sensing direction and of location $\mathbf{P_j}$, respectively. $\hat{D}_i^j$ is an estimate for $\mathbf{P_j}$ based on the ith direction sensing value. D_i^0 is the current measurement taken from $\mathbf{P_0}$ in the ith direction. Thus, the estimated proximity values for location $\mathbf{P_j}$ can be written as follows:

$$\hat{\mathcal{S}}_j = \left[\hat{D}_1^j,\ \hat{D}_2^j,\ \cdots,\ \hat{D}_i^j,\ \cdots,\ \hat{D}_{\mathcal{N}}^j\right]. \tag{9.13}$$

Figure 9.2 illustrates the proposed association scheme. From the figure, it is clear that an estimated proximity value in one of the $\mathcal{N}$ directions gives true proximity information if the sensing direction is perpendicular to the edge of a sensed object and the edge is long enough. Otherwise, the estimate provides an approximation; based on Figure 9.2, we can readily derive its error as follows:

$$e = \|\tilde{D}_i^j - \hat{D}_i^j\| \tag{9.14}$$

where $\tilde{D}_i^j$ denotes the true proximity value in the ith sensing direction at location $\mathbf{P_j}$.

It should be pointed out that the amount of errors induced in our association scheme depends on the complexity of the given environment. In order to express our confidence in accepting proximity estimates, we define here a confidence weight for each element of $\hat{\mathcal{S}}_j$, that is, a function of the distance between an agent and location $\mathbf{P_j}$, or, specifically,

$$w_j = e^{-\eta\rho_j^2} \tag{9.15}$$

where η is a positive constant and ρ_j is the distance, as shown in Figure 9.2.

According to Eq. 9.15, the weight is equal to 1 if the agent is located exactly at $\mathbf{P_j}$. That means $\hat{\mathcal{S}}_j|_{j=0}$ is the true value.

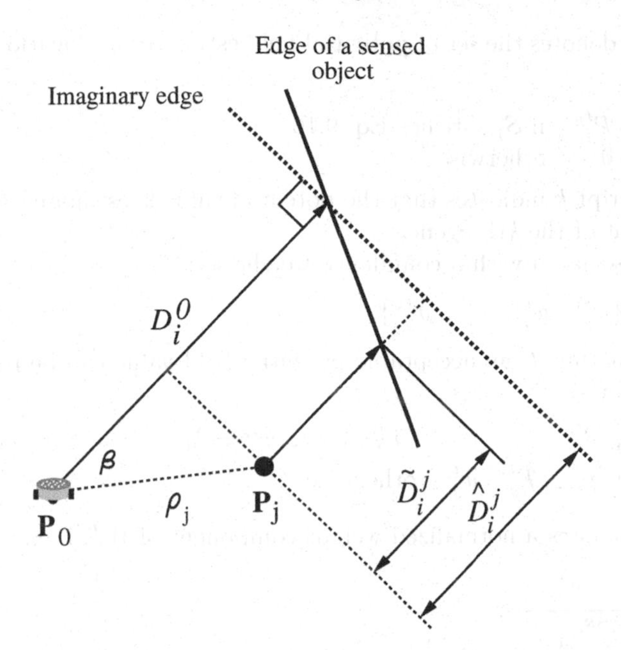

Fig. 9.2. An illustration of the proposed association scheme that computes proximity estimations in a surrounding region based on the proximity measurement obtained at the current location.

9.5.4 Incremental Self-Organization of a Potential Map

Considering a location in the environment at time t, its potential-field value can be calculated from its $\hat{\mathcal{S}}_j$ vector, as long as the vector satisfies the following condition:

$$\forall i \in [1, \mathcal{N}], \;\; \hat{D}_i^j \geq 0. \tag{9.16}$$

Location $\mathbf{P_j}$ is referred to as a *visible* location when the above condition is satisfied. The potential-field estimate at visible location $\mathbf{P_j}$ is computed as follows:

$$\hat{\mathcal{P}}_j^t = \sum_{i=1}^{\mathcal{N}} e^{-\lambda \hat{D}_i^j} \tag{9.17}$$

where λ is a positive constant. If at some locations Eq. 9.16 is not satisfied, the agent will simply discard those locations, as they are considered part of an object.

Thus, at time t, a set of potential-field estimates, $\Omega_t^j = \{\hat{\mathcal{P}}_j^{t_1}, \hat{\mathcal{P}}_j^{t_2}, \cdots, \hat{\mathcal{P}}_j^{t_k}\}$, can be derived by k agents with respect to location $\mathbf{P_j}$; that is,

$$\Omega_t^j \longleftarrow \Omega_{t-1}^j \bigcup \mathcal{Q} \tag{9.18}$$

where Ω_{t-1}^{j} denotes the set of potential-field estimates for location $\mathbf{P_j}$ at time $t-1$, and

$$\mathcal{Q} = \begin{cases} \hat{\mathcal{P}}_j^{t_k} & \text{if } \hat{\mathcal{S}}_j \text{ satisfies Eq. 9.16,} \\ \emptyset & \text{otherwise} \end{cases} \tag{9.19}$$

where subscript k indicates that the potential value is estimated based on the measurement of the kth agent.

Ω_t^j is associated with a confidence weight set:

$$W_t^j = \{w_j^{t_1}, w_j^{t_2}, \cdots, w_j^{t_k}\}. \tag{9.20}$$

Hence, at time t, an acceptable potential-field value can be readily calculated as follows:

$$\mathcal{P}_j^t = \begin{cases} \hat{\mathcal{P}}_j^{t_i} & \exists i \in [1, k]\,,\ w_j^{t_i} = 1, \\ \sum_{i=1}^{k} \hat{\mathcal{P}}_j^{t_i} \cdot \bar{w}_j^{t_i} & \text{otherwise} \end{cases} \tag{9.21}$$

where $\bar{w}_j^{t_i}$ denotes a normalized weight component of W_t^j, i. e.,

$$\bar{w}_j^{t_i} = \frac{w_j^{t_i}}{\sum_{n=1}^{k} w_j^{t_n}}. \tag{9.22}$$

The preceding Eqs. 9.18 and 9.21 are referred to as the *incremental self-organization* of a potential-field value. Assume that $\mathbf{M}$ agents are working simultaneously in the environment. At time t, after their distributed sensing, association, and potential-field self-organization, a global potential-field map covering all locations can be obtained.

9.6 Experiments

In order to experimentally validate our proposed potential-field-map-building approach, we place a group of mobile agents into a two-dimensional environment to perform the task as mentioned in the preceding sections. Figure 9.3 shows the experimental environment and the initial spatial distribution of the autonomous agents, where symbol '$*$' denotes the location of a single mobile agent. The environment contains four (4) stationary objects. All the agents are homed at one of the corners in the unknown environment. At the beginning, the agents have no *a priori* knowledge about their reactive behaviors.

The parameters used in this experiment are given in Table 9.1. In the table, population size p and generation size $\mathcal{G}$ are chosen according to the number of group agents, m, involved in the evolution; they are listed with respect to a group of size 1 to 6. As the generations of reactive motion behaviors evolve, the probability of mutation will decrease in a step-by-step fashion.

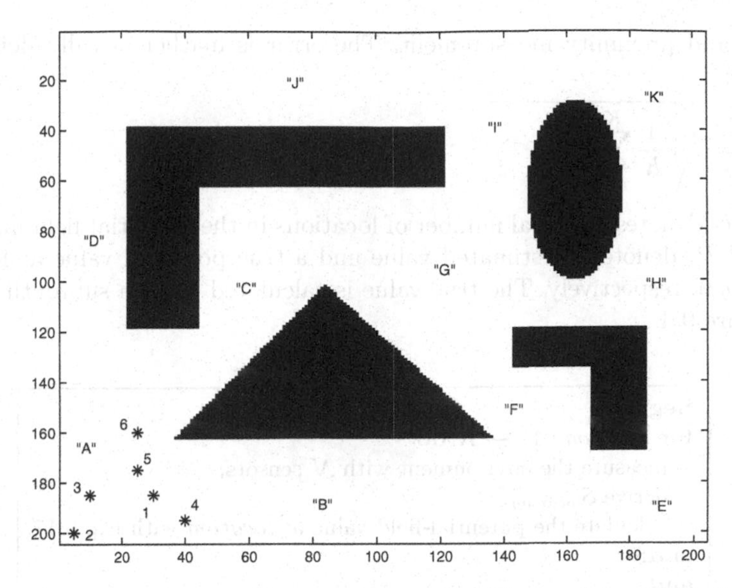

Fig. 9.3. The environment used for experimentation. The numbers signify the initial locations of the group agents, and the capital letters are the regional labels that will be referred to in later discussions.

Table 9.1. The parameters as used for the experiments.

parameter	symbol	unit	value
number of agents	$\mathbf{M}$		6
sensory section	$\mathcal{N}$		16
environment size		*grid* × *grid*	204×204
map resolution		*grid*	5
map locations			40×40
max. step increment	d_{max}	*locations*	7
constant	η		$\frac{1}{600}$
constant	λ		$\frac{1}{5}$
behavioral vector increment	ψ		0.2
chromosome length (per agent)	$2\mathcal{L}$	*bits*	8
population size	p		20/30/45/65/90/120
generations per step	$\mathcal{G}$		8/12/18/26/36/48
crossover probability	c		0.6
mutation probability	p_m		0.1/0.05/0.005
distance threshold	T_1	*grid*	10
distance threshold	T_2	*grid*	15

9.6.1 Experimental Design

We will compute the second-moment errors for all locations in a derived potential-field map with respect to a true map at each step of behavior evolution and proximity measurement. The error is mathematically defined as follows:

$$\varepsilon^t = \sqrt{\frac{1}{K}\sum_{j=1}^{K}(\mathcal{P}_j^t - \bar{\mathcal{P}}_j)^2} \tag{9.23}$$

where K denotes the total number of locations in the potential-field map, and $\mathcal{P}_j^t$ and $\bar{\mathcal{P}}_j$ denote an estimated value and a true potential value at location $\mathbf{P_j}(x_j, y_j)$, respectively. The true value is calculated using a subroutine given in Figure 9.4.

begin
for *location* : $1 \longrightarrow K$ **do**
  measure the environment with $\mathcal{N}$ sensors,
  derive $\mathcal{S}_{location}$,
  calculate the potential-field value at *location* with Eq. 9.17.
endfor
end

Fig. 9.4. The calculation of a true potential-field map.

9.6.2 Comparison with a Non-Adaptive Mode

In order to evaluate the performance of our proposed approach to incremental potential-field map building, we will also compare its resulting map with the one generated by a group of autonomous agents that utilize some local reactive motion behaviors, but without any behavior evolution (i.e., a non-adaptive mode). In both adaptive and non-adaptive modes, we will use the same set of parameters as in Table 9.1, including the number of group agents, their initial spatial distribution, and their maximal movement step increment.

In the non-adaptive mode, individual agents select and execute their local motions by calculating the standard deviation of potential-field values within their neighboring regions of radius d_{max} (i.e., the maximal movement step increment). Each new movement at time $t+1$ will be determined by two things, namely motion direction ξ and step increment s. Thus, a location at time $t+1$, $\mathbf{P_0^{t+1}}$, can be described as follows:

$$\mathbf{P_0^{t+1}} = \mathbf{P_0^t} + s \cdot e^{j\xi}. \tag{9.24}$$

More specifically, a vector Δ will be defined to record the standard deviation of potential value differences between times $t-1$ and t for all *sensing sectors*. The vth component of this vector is calculated as follows:

$$\Delta_v = \mathbf{std}(\{\delta_j \mid \delta_j = \mathcal{P}_j^t - \mathcal{P}_j^{t-1}, \;\; \forall j \in \mathcal{E}_v\}),\; v \in [1, \mathcal{N}] \tag{9.25}$$

where the operator **std** denotes the standard deviation for all locations inside the vth sensing sector, $\mathcal{E}_v$.

In addition, another vector Λ will also be defined to store the standard deviation of potential values at time t around each *location* within the same sensing sector. The jth component of this vector is determined as follows:

$$\Lambda_j = \mathbf{std}(\{\mathcal{P}_{j,l}^t \mid \forall j \in \mathcal{E}_v, \ |j-l| \leq 1\}) \tag{9.26}$$

where $\mathcal{E}_v$ denotes a sector selected as in Eq. 9.25.

Having made the above computations, agent i in a non-adaptive mode will choose its next movement direction θ_i^v (i.e., the vth sensing sector) whenever the following are satisfied:

$$\theta_i^v|_{\Delta_v} = \max(\Delta_1, \Delta_2, \cdots, \Delta_N) \tag{9.27}$$

and

$$\forall k \in [1, M] \text{ and } k \neq i, \quad P_k \notin \mathcal{E}_v \tag{9.28}$$

where the operator max returns the maximum from a set of values. P_k denotes the position of agent k.

Having determined its movement direction sector, the agent will further choose its next location $\mathbf{P}_0^{t+1}(x_0, y_0)$ to move to within the chosen sector. This location is the one that satisfies the following condition:

$$(x_0, y_0)|_{\Lambda_j(x_0,y_0)} = \mathbf{max}(\Lambda_1, \Lambda_2, \cdots). \tag{9.29}$$

9.6.3 Experimental Results and Comparisons

Figure 9.5 presents a potential-field map obtained in the above-mentioned adaptive map-building mode, along with its contour plot. It corresponds to the exploration in the unknown environment for 20 movement steps by six autonomous agents in a genetic algorithm (GA)-based adaptive mode.

In order to examine the overall spatial distribution of group agents during map building, Figures 9.6 and 9.7 present the locations in the unknown environment visited by the agents in two different modes, respectively. It is interesting to observe that the group agents with behavior evolution capabilities can in general sample the unknown environment evenly, along the valley of the potential field. They will focus slightly more on the junction locations in order to eliminate the proximity uncertainty involved. In the non-adaptive mode, however, the agents quite often go to the locations near the edges of the objects where large standard deviations of the map can be found.

In addition to the above observations, Figure 9.8 further presents a second-moment error comparison between the two modes of map-building. It is obvious from the figure that the error in the case of evolutionary agents goes down much more quickly than in the non-adaptive case.

In the adaptive mode, the agents are dynamically distributed to collectively build a potential-field map in about ten movement steps. Their navigation tends to explore uncovered areas first, and then tries to refine local

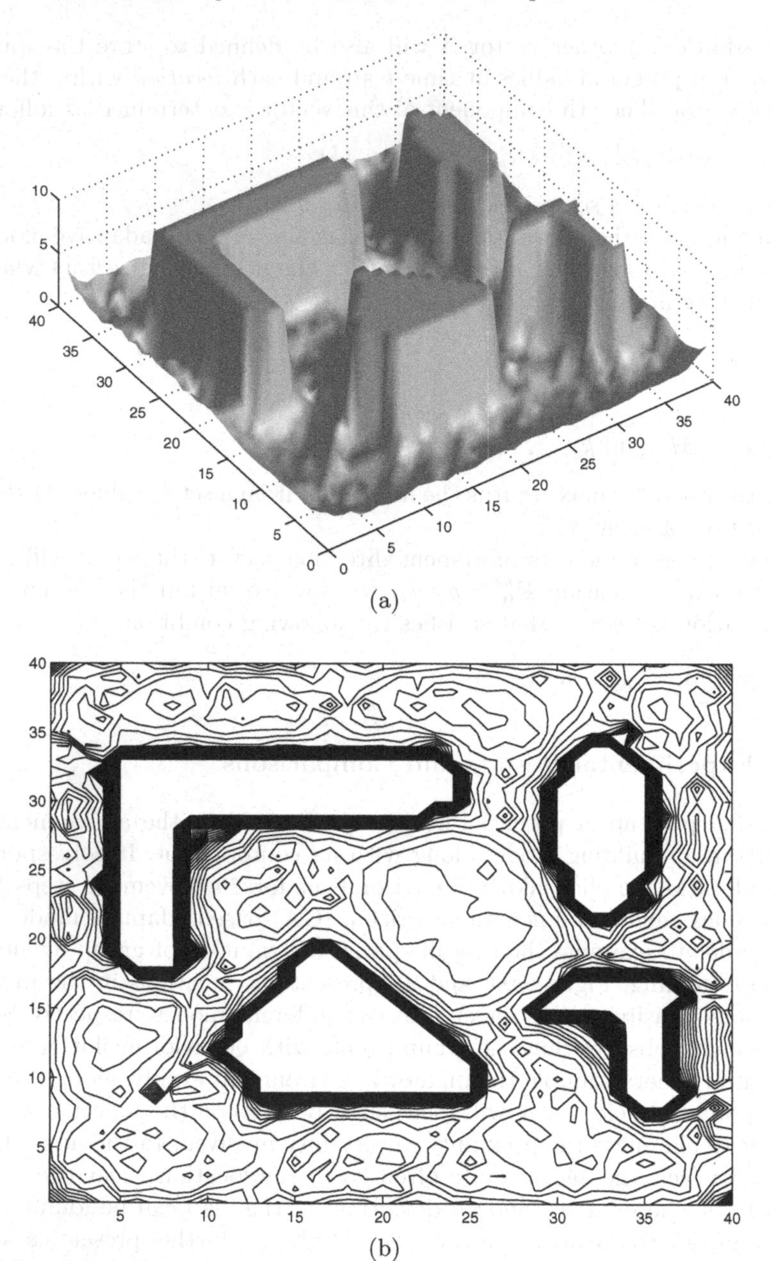

Fig. 9.5. Adaptive mode: (a) A map built by six cooperative autonomous agents after 20 movement steps. (b) A contour plot for the obtained potential-field map.

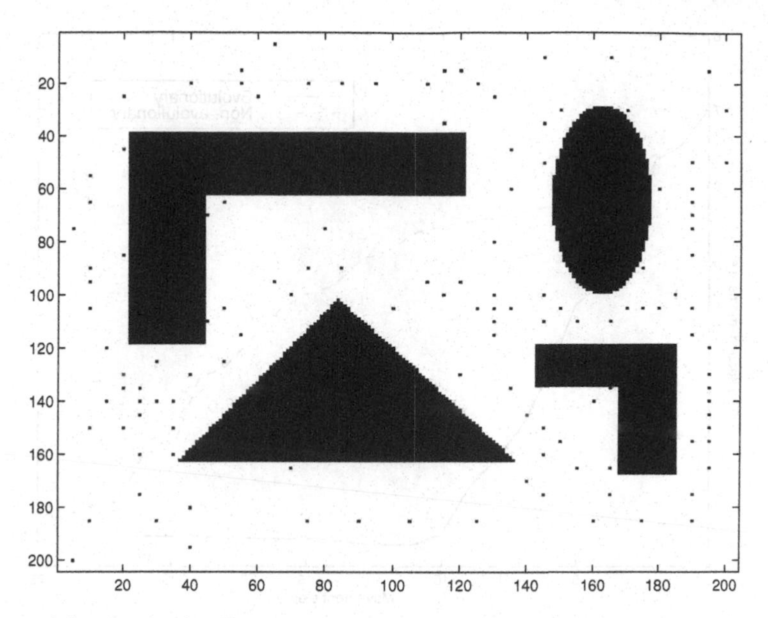

Fig. 9.6. Adaptive mode: Locations visited by a group of evolutionary autonomous agents.

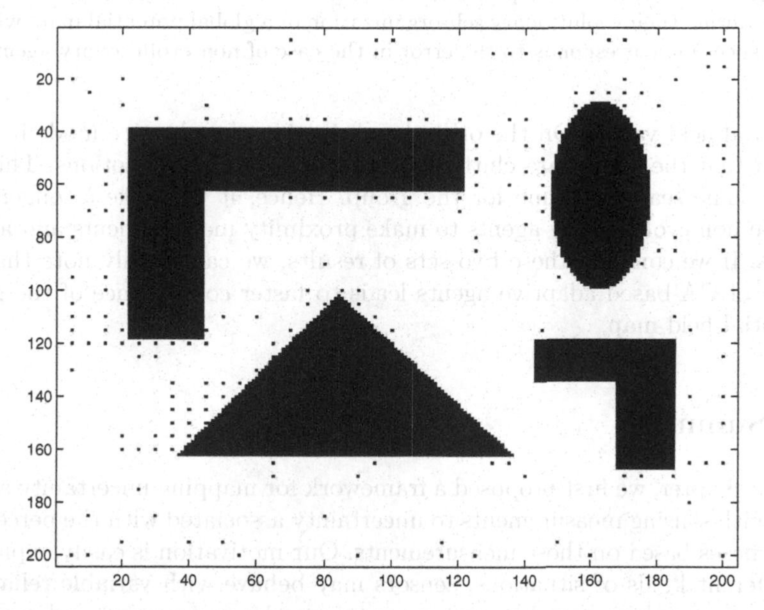

Fig. 9.7. Non-adaptive mode: Locations visited by a group of non-evolutionary agents.

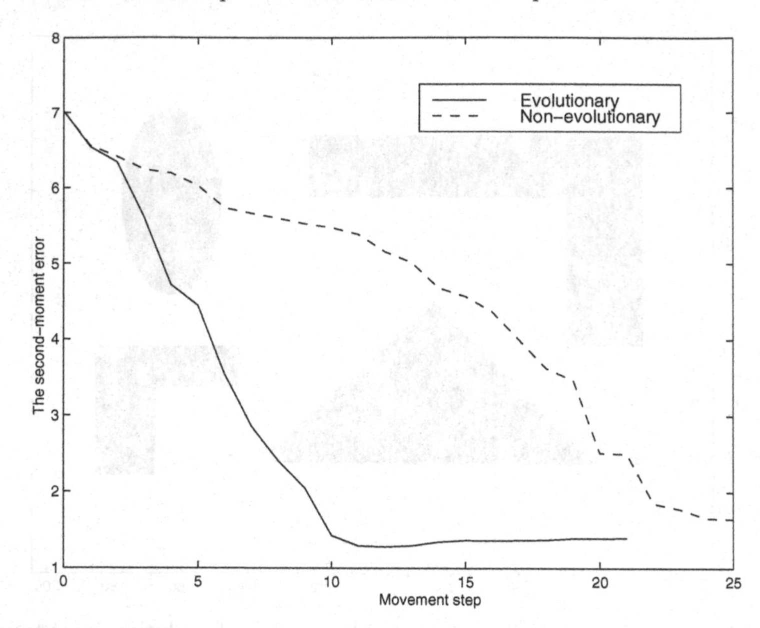

Fig. 9.8. A second-moment error comparison between the two modes of map-building. The solid line corresponds to the error measured in the case of evolutionary agents during their evolutionary self-organization of a global potential map, whereas the dashed line corresponds to the error in the case of non-evolutionary agents.

potential-field values. On the other hand, in the non-adaptive mode it is observed that the agents are cluttered together during their motions. This will waste some searching time for the group. Hence, it will take a longer time for the non-evolutionary agents to make proximity measurements and associations. If we compare these two sets of results, we can readily note that the group of GA-based adaptive agents leads to faster convergence of the global potential-field map.

9.7 Summary

In this chapter, we first proposed a framework for mapping uncertainty associated with sensing measurements to uncertainty associated with the perceptual hypotheses based on those measurements. Our motivation is easily explained: in different kinds of situations, sensors may behave with variable reliability, i.e., sensing values may be more uncertain or less uncertain, and such uncertainty must be taken into account when deciding how to 'react' to the current perceptual hypothesis. For the case of a mobile robot, this may mean 'trusting' the hypothesis "no obstacle ahead" differently in different kinds of situations. If there is much uncertainty associated with the hypothesis, then the robot may decide to advance or to turn slightly in position and take another measurement instead of moving forward. Our treatment is mathematically correct, practically relevant, and experimentally functional, at least in an idealized world.

Later in this chapter, we described a GA-based evolutionary computation approach to the emergence of cooperative group behaviors for incrementally self-organizing a global potential-field representation in an unknown environment. While giving the underlying modeling and computation formalisms, we presented several results from our experimental validation. Generally speaking, our approach enables the distributed agents to gradually develop an ability of experience-based cooperation and adaptation to their task environment.

10

Concluding Remarks

This book has presented a unified approach to analyzing and planning the configurations of planar mechanisms. The approach utilizes a qualitative spatial representation and reasoning framework integrated with a quantitative search procedure. Algorithms for mechanism analysis and motion planning have been described in the contexts of constrained closed-chain linkage mechanisms and under-constrained open-chain mechanisms. The proposed algorithms have been validated with simulations based on implemented programs.

10.1 Key Concepts Revisited

The work described in this book contributes to the understanding of spatial reasoning and planning as well as its applications to planar mechanism analysis and motion planning. The specific contributions can be highlighted as follows:

1. A *framework* of qualitative spatial representation and reasoning has been systematically constructed. This framework can be summarized in the following steps:
   a) Define the qualitative abstractions of quantitative spatial variables;
   b) Define domain-specific *constraints* (e.g., inference rules or weighted connectivity graphs) for the search of spatial descriptions (e.g., a polygonal configuration or a connected route);
   c) To generate exact spatial relationships, (i) find the qualitative description of the solution by means of constraint propagation and (ii) apply standard local optimization techniques to search for an exact quantitative solution.
2. Formalisms for the qualitative abstraction of spatial quantities, such as the Euclidean *distances* and *angles*, were constructed, and accordingly spatial inference rules, called *qualitative trigonometry* ($\mathcal{QT}$) and *qualitative arithmetic* ($\mathcal{QA}$), were formulated. Theorems concerning the minimum requirements and completeness of the spatial inferences (based on the $\mathcal{QT}$ and $\mathcal{QA}$ rules) were developed.

3. A formalism for the spatial characterization of a polygonal environment based on the notion of *m-closure regions* was introduced. Several properties of the m-closure partition of the environment were identified, with respect to the uniqueness and upper and lower bounds of the regions. Based on this representation, the notions of *qualitative locations* and *qualitative routes*, as well as three qualitative *optimality criteria* for measuring the length, passage clearance, and orientation cost of the route were defined. Based on the definition of m-edge configurations, a new segmentation scheme was developed to represent a path within a polygonal environment.
4. The qualitative spatial inferencing technique was applied to deriving the instantaneous configurations as well as velocity relationships of *planar linkages.* A set of detailed algorithms for controlling the spatial propagation was formulated. The qualitative approach produces the ranges of spatial variables that can limit the possible configurations for local search (i.e., search space for simulated annealing) in generating exact configurations. This approach is suited for problems where the initial information about the metrics of the linkages is given in *qualitative* terms.
5. The technique for planning a global qualitative route as well as an exact path given a set of connected m-closure regions, was demonstrated with *robot-like under-constrained mechanisms* in the presence of a set of static obstacles. The local exact path-segment search given a route description was implemented using simulated annealing.

10.2 Practical Application

10.2.1 Computer-Aided Mechanism Analysis

In kinematic analysis, it is always desirable to know what motions the various parts of a mechanism approximately undergo, and the relationships between these motions. In classical kinematics, such an analysis usually requires *explicit quantitative information.* The problems associated with quantitative approach are (1) that generating a solution usually constitutes a major computing task and (2) that the generated solution has to be carefully interpreted by humans if the function of the mechanism is to be understood.

In order to build intelligent systems that can perform effective reasoning about the function of a mechanism based on incomplete specifications and can communicate the results with users at a functional and qualitative level, alternative kinematics frameworks have to be developed.

In this context, our book has focused on the qualitative approach to kinematics. It offers a set of specific solutions as to how qualitative geometric reasoning can be applied to solve kinematic analysis problems. These solutions will not only serve as a useful framework for kinematics, but will also evolve and necessitate successful application of Artificial Intelligence (AI) technology in applied kinematic analysis.

10.2.2 Robot Compliant Task Analysis

In real-life situations, the precise geometry of a mechanism to be manipulated may not be available *a priori* knowledge due to sensing or modeling limitations [17]. What one can expect, though, is to gather and fuse low-level sensory data from some external sensors (*e.g.*, TV cameras) into a *qualitative* representation of the mechanism geometry. In planning mechanical device manipulation tasks with uncertainty, the current constrained mechanism analysis approach may be useful in the following ways:

1. *The behavior or function of a device is to be achieved through operating some of its parts.* Given a description of a device configuration in terms of its parts and their topological connections, a task planner must determine what motions various parts undergo if a particular part is being manipulated and how the motion variations in one part functionally affect the others. Modeling and simulation of the kinematic state transitions in devices enable the determination of possible robot operations.
2. *Compliant motion strategies are to be specified with respect to a compliance frame.* Robot actions that are difficult to express in the world frame may simply be specified with respect to a compliance frame attached (for the duration of the task) to the part being manipulated. In such a case, the trajectory of the compliant frame with respect to a world model needs to be computed. This is a non-trivial problem, since in mechanisms the constrained trajectories at different points of the same link may differ.
3. *Compliant motion strategies are to be specified with respect to a world frame.* In this case, kinematic constraints of the parts being manipulated act to create a pattern of forces that the feedback strategies can sense and then use to modify the robot motions.

10.2.3 Robot Path Planning

The path-planning approach for under-constrained mechanisms, as presented in this book, is adequately general, and can be applied to planning either single mobile robots or manipulators of many degrees of freedom.

This approach is designed to find near-optimal paths, and is best suited to situations where either the exact robot environment is known or the model of the environment contains some uncertainty. However, if the knowledge about the robot and its environment is qualitative, the route planning phase will still be applicable. Such qualitative routes may be used to guide other local or reactive planners.

The *configuration-space-based algorithms* [56, 140] require both a memory space and computation time exponential in number of configuration parameters. This has imposed a certain limitation on the number of degrees of freedom a manipulator may have.

The efficiency of the current path-planning approach for under-constrained mechanisms is, in part, due to the fact that the path segments between m-edge configurations are not searched online, but instead computed offline and retrieved as necessary. This significantly reduces the computational cost compared to the previously developed approaches.

Even for the offline computation, the current approach shows a considerable efficiency, since it checks whether or not a given configuration is collision-free only with respect to the objects that are connected to the current *m*-closure regions. Therefore, this offline process may also be applied in an online process. Furthermore, if the dedicated hardware with parallel computing power is created for the annealing process, a real-time planner may be feasible.

One of the drawbacks of the previous *pure local planning methods* is that the methods rely on the *minimization* of a function such as a distance. In the current approach, the qualitative routes serve as a global plan and the random motions are performed only between adjacent regions; therefore, the robot will not likely get trapped in dead-ends (i.e., local minima).

10.3 Limitations

The proposed approach is subject to some limitations, as have already been discussed in Sections 6.4.2 and 8.5.5. Generally speaking, these limitations can be summarized as follows:

1. The path derivation process is *not deterministic.*
2. Exact paths contain high spatial frequencies (i.e., *not smooth*).
3. *No precisely optimal paths* will be found for under-constrained mechanisms.
4. *Not all* existing paths are guaranteed to be found.

10.4 Future Challenges

This book has presented several opportunities for future research. In the longer term, since the direct implications of the current work are on computer-aided mechanism analysis for post-conceptual design and robot (compliant) motion planning, the practical application in some of these areas, as described earlier, remains to be explored. In the shorter term, several extensions from the current work should be further investigated. These extensions are described in the following sections.

10.4.1 Simulated Annealing

As mentioned in Chapters 6 and 8, the performance of the quantitative search (`ANNEAL`) varies with respect to the controlling parameters. The most predominant parameter is the *temperature.* Although the temperature function is likely to be problem-dependent, it seems interesting to study the effects of different functions and provide more insights into general adaptive mechanisms for temperature variation. One possible extension is to work on a temperature control mechanism based on an interactive process or a set of heuristic rules.

For the experiments conducted in this work, the probability distribution of annealing configuration generation is static. It will be more challenging to explore the feasibility of dynamic updating of the probability distribution with a *learning module* based on previous iterations of configuration test.

In addition to *simulated annealing*, other quantitative search techniques, such as *dynamic programming* and *genetic algorithms*, may also be worth more extensive investigation.

10.4.2 Local Path Planning near m-Edges

For all the experimental cases used in the current work, the proposed algorithms have sufficed to find near-optimal paths with respect to the given geometric information. However, improvements are still possible to make the algorithms more robust.

Yap [195] has proposed exact *local* path-planning algorithms for common stereotypes, such as door passing. These algorithms may be incorporated with the current path-planning approach to find local path segments *across m-edges*.

10.4.3 Other Heuristic Search Strategies for Qualitative Route Planning

In the present work, the search strategy used in the qualitative route planning is based on *hill-climbing*, a depth-first search technique with an ordering of choices. Another possible strategy *for the route planning* is the A^* search algorithm. This requires, to some extent, the modification of the cost functions. Appendix C provides a formulation of such a search strategy, directly extended from the current work.

10.4.4 Incorporating a Continuous Manipulator Model

Hayashi and Kuipers [81] have presented a novel path-planning approach for highly redundant manipulators using a *continuous model*. This approach uses cubic spirals to provide a continuous curvature path in the free-space, and thus it reduces the complexity of the planning problem, from a function of both the degrees of freedom of the manipulator (exponential) and the complexity of the environment (polynomial), to a polynomial function of the complexity of the environment only.

The path-planning technique for redundant manipulators developed in this book utilizes a qualitative route to guide the local search of exact path segments. Such a local search procedure may also be carried out by applying Hayashi and Kuipers's approach from one m-closure region to another, along the qualitative route. It will be of great interest to explore the possibility of planning the local continuous curvature path segments given only a qualitative spatial representation of the free-space.

10.4.5 Extensions to Complex Mechanisms and Non-Convex Obstacles

This study has concentrated on planar mechanisms with lower kinematic pairs. Therefore, an interesting extension will be to *generalize* the under-constrained mechanism analysis algorithms so that mechanisms with higher pairs can also be handled.

Another related issue is to extend the path-planning algorithms to include obstacles with concave shapes. Although in the simulations of the present work the tested cases have already involved non-convex environments, there is still a need for a formal representation of such an environment. One possibility will be to first partition each concave obstacle into a set of disjoint convex ones, and then to apply the current qualitative route planning and quantization techniques.

Broadly speaking, the mechanism analysis problems as investigated in this book may be generalized into motion planning within the following two types of environments:

1. **Strongly Connected Environment (SCE):** A set of moving and stationary rigid objects kinematically constrained by lower-pair joints.
2. **Weakly Connected Environment (WCE):** A set of moving and stationary rigid bodies spatially constrained through a combination of the following actual or virtual contacts:
   a) point contact;
   b) line or curve contact; and
   c) virtual contact (i.e., the distance between two bodies is greater than 0 but less than a predefined threshold).

These types of environments may be treated in a unified framework. Some interesting ideas have been proposed previously [134], where a set of motion primitives (e.g., *move* and *insert*) are formulated and can be syntactically manipulated.

Finally, the qualitative treatment of the mechanism motion planning should include reasoning about the *dynamics* of the physical world. A starting point will be to extend the current qualitative framework to reason about the dynamics of planar mechanical devices under robot manipulation. In qualitative dynamics, force propagation must be analyzed in addition to motion propagation.

A

Grid-Based Representation for Totally Ordered $\mathcal{Q}$-space

The quantity spaces for qualitative distances and qualitative angles, as described in Sections 4.1 and 10, are totally ordered, since for every $x, y \in \mathcal{Q}$-space, either $x \leq y$ or $y \leq x$ is true.

To represent a totally ordered *two*-dimensional $\mathcal{Q}$-space, a rectangular-cell-based, *two*-dimensional grid structure, called *qualitative grid*, may be used. In the case of a joint space of $\mathcal{Q}$-space$_{[l]}$ and $\mathcal{Q}$-space$_{[\theta]}$, the values of *one*-dimensional length $\mathcal{Q}$-space, $[l]_i$, are represented as rows within a *single column* and the values of *one*-dimensional angle $\mathcal{Q}$-space, $[\theta]_j$, are represented as columns within a *single row*. In the qualitative grid, two elements are said *adjacent* if they are immediate neighbors of each other.

Based on the above graphical convention, the $\mathcal{QT}$ and $\mathcal{QA}$ rules can be represented using qualitative grids, as shown in Figures A.1 and A.2, respectively.

Freksa [68] has previously used a slightly different qualitative grid representation, called an *iconic* composition table, to express the rules for inferring nine location and orientation relations. Freksa's composition table can be considered a simplified form of qualitative orientation theory, which can be logically deduced from the rules of Figure A.1.

As can be readily noted, in Figures A.1 and A.2, the larger the dark area, the greater the uncertainty. Furthermore, the cells (i.e., qualitative values) within each entry are connected, and that the cells in one entry are either overlapping with or adjacent to the cells in its neighboring entries. These observations are consistent with Theorem 8.

With such a grid-based representation, the example of qualitative configuration analysis, as described in Section 6.2.1, can be readily understood and implemented. Figure A.3 presents a grid representation of the spatial relations given in this example. Figure A.4 graphically summarizes the entire process of the qualitative configuration analysis for the four-bar linkage shown in Figure 6.2.

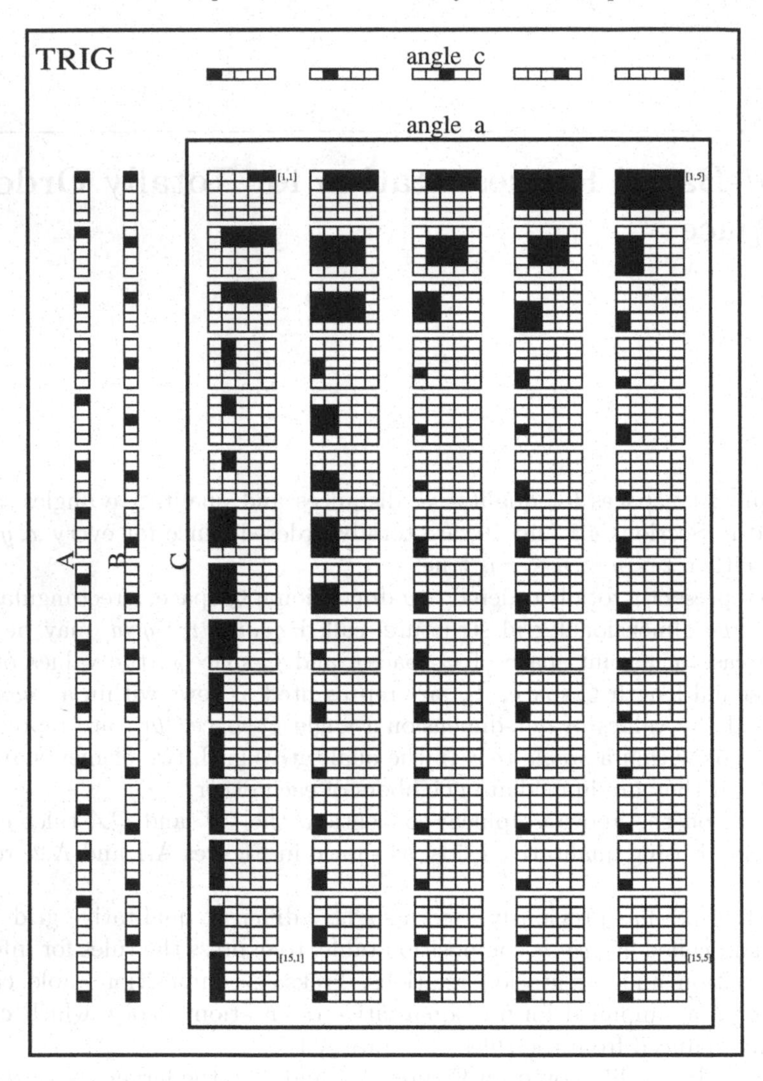

Fig. A.1. Qualitative spatial relationships of planar triangles.

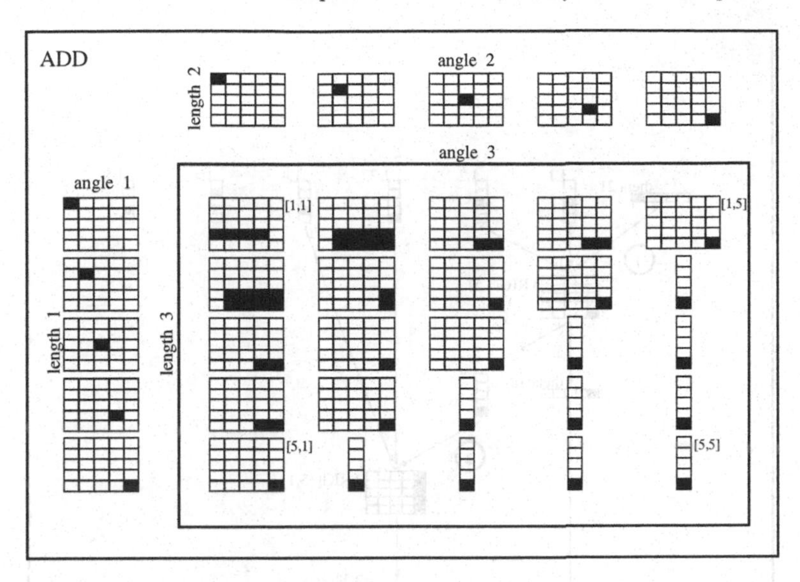

Fig. A.2. Qualitative addition operation.

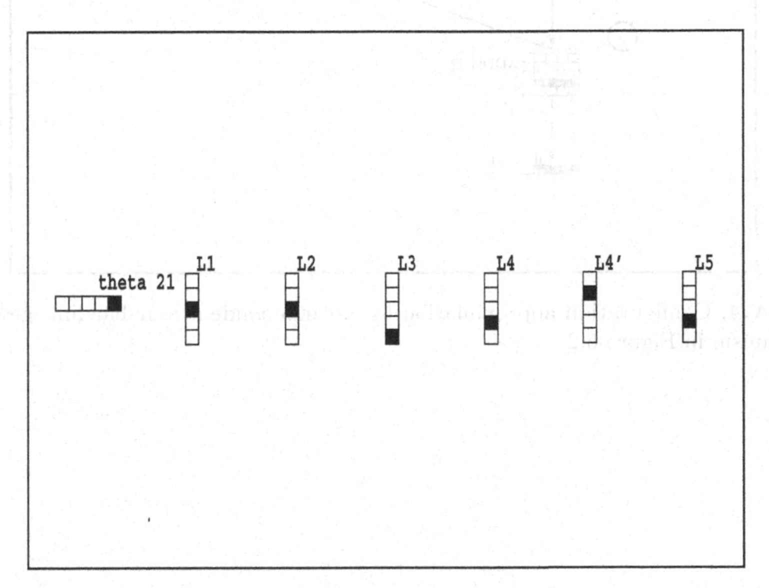

Fig. A.3. A grid representation of the given qualitative spatial relationships for the constrained mechanism shown in Figure 6.2.

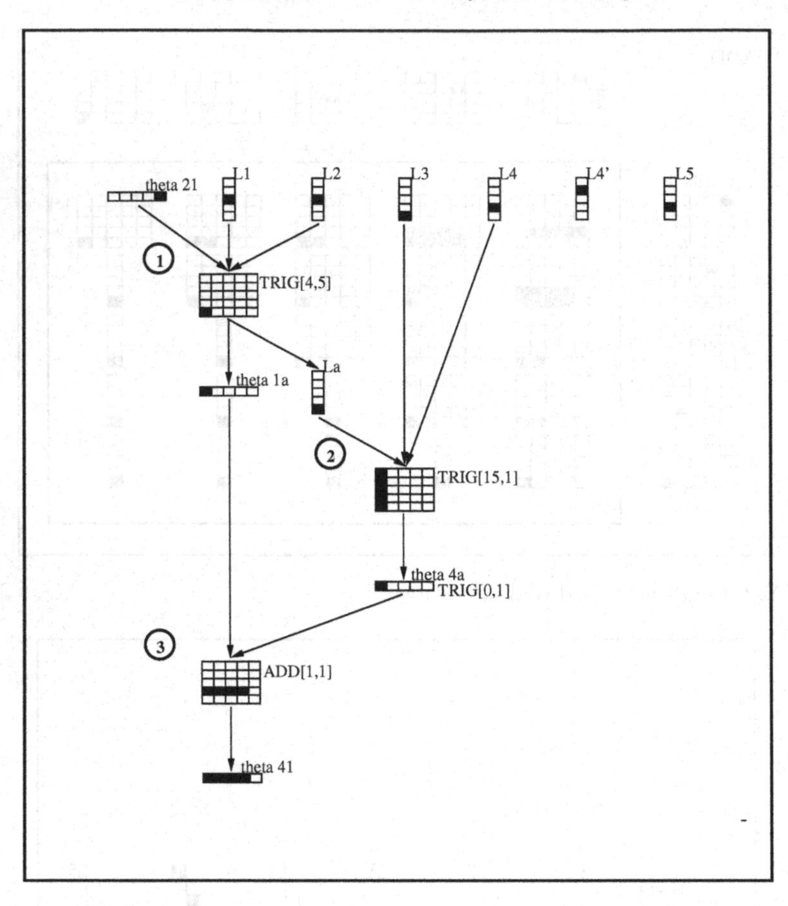

Fig. A.4. Configuration approximation for an independent four-bar linkage of the mechanism in Figure 6.2.

B

The Boltzmann Distribution in Simulated Annealing

This appendix provides one explanation of the probability distribution as used in the `ANNEAL` algorithm. Its more detailed discussion can be found in [174].

Suppose that there is a system which randomly generates $\{\mathbf{C}_1, \mathbf{C}_2, ..., \mathbf{C}_k, ..., \mathbf{C}_\mathrm{N}\}$, a set of N neighboring states of $\mathbf{C}_\mathrm{init}$. Each state can be specified by a set of parameters, i.e., $\mathbf{C}_k = (x_1, x_2, ..., x_i, ..., x_m)$. The neighboring states can be considered as the results of randomly varying individual parameters to their neighbors with equal probabilities. The distance between states $\mathbf{C}_k$ and $\mathbf{C}_\mathrm{goal}$ is measured by their Euclidean distance, i.e., $d_\mathrm{k,goal} = |\mathbf{C}_k - \mathbf{C}_\mathrm{goal}|$.

The total of the distances from N neighboring states to the goal configuration can be written as

$$D = \sum_{s=1}^{t} N_s d_\mathrm{s,goal} \tag{B.1}$$

where N_s is the number of the states of the same distance $d_\mathrm{s,goal}$, s is the set of states with the sth shortest distance, and $t \leq N$.

Given N states, the possible ways of arranging a set of distinct states (in terms of distance) $\{N_1, N_2, ..., N_s, ...\}$ are given as follows:

$$W(N_1, N_2, ..., N_s, ...) = \frac{N!}{N_1!N_2!...N_s!...} \tag{B.2}$$

From practical experience, it is known that a large system of this kind will spontaneously approach an equilibrium state irrespective of the initial state. Therefore, it can be assumed that the system, after a large number of state changes, will reach an equilibrium state where the distribution of N states is $\{n_1, n_2, ..., n_s, ...\}$, ordered by corresponding Euclidean distances $l_s = d_\mathrm{s,goal}$. That is,

$$\frac{d}{dN_s} W(n_1, n_2, ..., n_s, ...) = 0, \quad \forall s \tag{B.3}$$

where $\sum_s n_s = N$.

It is assumed that N is a large number such that $N!$ can be approximated by Stirling's relation, $\ln N! \simeq N \ln N - N$.

This approximation leads to the following:

$$\ln W \simeq N \ln N - \sum n_s \ln n_s. \tag{B.4}$$

Hence, Eq. B.2 can be written as

$$d(\ln W) = -\sum_s \ln n_s dn_s = 0 \tag{B.5}$$

Similarly, one has

$$d\left(\sum_s n_s\right) = dn_s = 0 \tag{B.6}$$

and

$$d\left(\sum_s l_s n_s\right) = \sum_s l_s dn_s = 0. \tag{B.7}$$

Using Lagrange's multipliers α and β, the following can be obtained:

$$\ln n_s dn_s + \alpha dn_s + \beta l_s dn_s = 0. \tag{B.8}$$

Equivalently,

$$\ln n_s + \alpha + \beta l_s = 0. \tag{B.9}$$

Determining α from Eq. B.5, the distribution of n_s is found to be the Boltzmann distribution, as follows:

$$\frac{n_s}{N} = c \exp(-\beta l_s) \tag{B.10}$$

where the normalization constant

$$c = \frac{1}{\sum_s \exp(-\beta l_s)} \tag{B.11}$$

What this distribution says is that given a β, the system in equilibrium has a larger chance to find states which have shorter distances. At the same time, it also has a chance, albeit very small, to generate a new state from the previous state that increases the Euclidean distance. This will give the system an equal chance to get out of a local minimum in favor of finding a better one which has a globally shorter distance. Hopefully, the system will reach, or be very close to, a global minimum.

C

Qualitative Route Search Based on A^* Algorithm

Having constructed a weighted connectivity graph $\mathcal{G}$ associated with the m-closure partition of a given polygonal environment, the A^* algorithm [191] may be applied to find a route (i.e., a sequence of adjacent nodes), from given initial and goal regions, whose total cost is optimized with respect to some qualitative cost functions.

As mentioned in Section 4.7, each arc A in the graph is associated with a function c which returns a pair of qualitative values corresponding to the qualitative relative distance and the qualitative orientation. In A^* search, the cost of a route can be defined as the sum of the *costs of individual connecting arcs.*

The *qualitative optimality* of a path may either be specified with respect to a single cost function (e.g., minimum total length) or be defined with a joint function composed of various relatively weighted costs. In the latter case, a simple *arithmetic scheme* can be used to compute the total cost contributed by each weighted cost function. The arithmetic scheme is similar to the idea of arc counting in computing the length of a certain path. It counts the number of grid changes in calculating $[Orientation\_cost(i)]$, denoted by $\triangle J_{[\text{Orientation\_cost}]}$ where J represents the numerical index of a relative angle. In calculating $[Clearance(i)]$, denoted by $I_{[\text{Clearance}]}$, it directly uses the grid numerical index. An example of the composite cost function is given as follows:

$$C_{[\text{cost(k1,k2,...,ki)}]} = C_{[\text{cost(k1,k2,...,ki}-1)]} - \alpha I_{[\text{Clearance}]} + \beta \triangle J_{[\text{Orientation\_cost}]} + \gamma [Length\_inc] \quad \text{(C.1)}$$

where α, β, and γ are the coefficients that weigh the relative significance of a criterion.

During the search, heuristic information is incorporated into a qualitative evaluation function $[f_i^*]$ which is the sum of two functions $[g_i^*]$ and $[h_i^*]$. $[g_i^*]$ represents the qualitative cost from the initial m-closure region m_{init} to another region m_{i} through a sequence of connected m-edge crossing, and $[h_i^*]$ represents an estimate of the costs from region m_{i} to the goal region m_{goal}.

Formally, the A^* search algorithm can be stated as follows:

Algorithm QUALITATIVE_ROUTE

1. Mark the initial region m_{init} OPEN and calculate an evaluation function $[f^*_{\mathrm{m_{init}}}]$.
2. Select the open region m_i whose $[f^*_{\mathrm{m_i}}]$ value is the smallest; if ties occur, resolve in favor of a goal region.
3. If m_i is a goal region, mark m_i CLOSED and terminate the algorithm.
4. Otherwise, mark m_i CLOSED, calculate $[f^*]$ for each connected region of m_i, and mark OPEN each connected region not already marked CLOSED; remark as OPEN any CLOSED region m_j which is a successor of m_i for which $[f^*_{\mathrm{m_j}}]$ is smaller now than it was when m_j was marked CLOSED, and go to Step 2.

$[f^*_{\mathrm{m_i}}]$ is computed by using a simple arithmetic scheme incorporating relatively weighted costs, as mentioned above.

It should be noted that, unlike the A^* search of traditional graphs with explicit quantitative costs, the current application of the algorithm is not intended to generate an optimal path in a quantitative sense. Instead, this search procedure is used to derive a qualitative characterization of the routes in which the actual optimal paths are likely to be found.

References

1. N. Abe, S. Sako, and S. Tsuji. High-level robot task specification. In *Proceedings of the International Workshop on Artificial Intelligence for Industrial Applications*, pages 341–346, 1988.
2. J. F. Allen. Maintaining knowledge about temporal intervals. *Communications of the ACM*, 26(11):832–843, 1983.
3. R. Arnheim. *Art and Visual Perception*. University of California Press, Berkeley, CA, 1974.
4. I. Artobolevsky. *Mechanisms in Modern Engineering Design*, volume 1–4. MIR Publishers, Moscow. English Translation, 1979.
5. A. Baader and G. Hirzinger. A self-organizing algorithm for multisensory surface reconstruction. In *Proceedings of the 1994 IEEE/RSJ International Conference on Intelligent Robots and Systems*, pages 81–88, Munchen, Germany, Sept. 1994.
6. C. Bard and J. Troccaz. Automatic preshaping for a dextrous hand from a simple description of objects. In *Proceedings of the IEEE International Workshop on Intelligent Robots and Systems*, pages 865–872, 1990.
7. J. Barraquand and J.-C. Latombe. A monte-carlo algorithm for path planning with many degrees of freedom. In *Proceedings of the IEEE International Conference on Robotics and Automation*, pages 1712–1717, 1990.
8. C. Bellier, C. Laugier, E. Mazer, and J. Troccaz. Planning/executing six d.o.f. robot motions in complex environments. In *Proceedings of the 1991 IEEE/RSJ International Workshop on Intelligent Robots and Systems (IROS '91)*, pages 91–96, Osaka, Japan, Nov. 3–5 1991.
9. J. Bernasconi. Low autocorrelation binary sequences: statistical mechanics and configuration space analysis. *Journal of Physique*, 48(4):559–567, 1987.
10. S. Betge-Brezetz, P. Hebert, R. Chatila, and M. Devy. Uncertain map making in natural environments. In *Proceedings of the IEEE International Conference on Robotics and Automation*, pages 1048–1053, Minneapolis, MN, 1996.
11. M. Betke, R. L. Rivest, and M. Singh. Piecemeal Learning of an Unknown Environment. Memo 1474, MIT AI Lab, 1994.
12. A. Blackwell. Qualitative geometric reasoning using a partial distance ordering. In J. S. Gero and R. Stanton, editors, *Artificial Intelligence Developments and Applications*, pages 217–229. Elsevier Science Publishers B. V., North-Holland, 1988.

13. D. G. Bobrow. *Qualitative Reasoning about Physical Systems.* The MIT Press, Cambridge, MA, 1985.
14. D. G. Bobrow. Special volume on artificial intelligence in perspective. *Artificial Intelligence*, 59(1–2), 1993.
15. P. P. Bonissone and K. P. Valavanis. A comparative study of different approaches to qualitative physics theories. In *Proceedings of the Second IEEE Conference on AI Applications*, 1985.
16. T. Boseniuk, W. Ebeling, A. Engel, J. Boltzmann, and R. Darwin. Strategies in complex optimization. *Physics Letters*, 125(6–7):307–310, 1987.
17. M. Brady. The problems of robotics. In M. Brady, editor, *Robotics Science*, System Development Foundation Benchmark Series, pages 1–35. The MIT Press, Cambridge, MA, 1989.
18. R. A. Brooks. Solving the find-path problem by good representation of free-space. *IEEE Transactions on Systems, Man, and Cybernetics*, 13(3):190–197, 1983.
19. R. A. Brooks and T. Lozano-Pérez. A subdivision algorithm in configuration space for findpath with rotation. In *Proceedings of IJCAI-83*, pages 799–806, Aug. 1983.
20. R. Brost. Computing metric and topological properties of configuration-space obstacles. In *Proceedings of the 1989 IEEE International Conference on Robotics and Automation*, pages 170–176, 1989.
21. S. J. Buckley. Planning compliant motion strategies. *The International Journal of Robotics Research*, 8(5):28–44, Oct. 1989.
22. S. P. Carney and D. C. Brown. A continued investigation into qualitative reasoning about shape and fit. *AI EDAM*, 3(2):85–110, 1989.
23. V. Cerný. Thermodynamical approach to the traveling salesman problem: an efficient simulation algorithm. *Journal of Optimization Theory and Applications*, 45(1):41–51, Jan., 1985.
24. B. Chandrasekaran. Design problem solving: a task analysis. *AI Magazine*, 11(4):59–71, 1990.
25. S. K. Chang and E. Jungert. A spatial knowledge structure for image information systems using symbolic projections. In *Proceedings of the National Computer Conference*, pages 79–86, Dallas, Texas, Nov 2–6, 1986.
26. B. Chazelle. Approximation and decomposition of shapes. In Jacob T. Schwartz and Chee-Keng Yap, editors, *Algorithmic and Geometric Aspects of Robotics*, volume 1, pages 145–185. Lawrence Erlbaum Associates, Publishers, Hillsdale, NJ, 1987.
27. H. Chu and H. A. ElMaraghy. Real-time multi-robot path planner based on a heuristic approach. In *Proceedings of the 1992 IEEE International Conference on Robotics and Automation*, pages 475–480, Nice, France, May 1992.
28. I. Cochin. *Analysis and Design of Dynamic Systems.* Harper and Row, New York, NY, 1980.
29. N. E. Collins, R. W. Eglese, and B. L. Golden. Simulated annealing bibliography. *American Journal of Math. Management Sci.*, 8(3 & 4):209–307, 1986.
30. P. Dague. Symbolic reasoning with relative orders of magnitude. In *Proceedings of IJCAI-93*, pages 1509–1514, Aug. 1993.
31. E. Davis. *Representing and Acquiring Geographic Knowledge.* Pitman Publishing, London, 1986.
32. E. Davis. Constraint propagation with interval labels. *Artificial Intelligence*, 32(3):281–331, 1987.

33. E. Davis. A logical framework for commonsense predictions of solid object behavior. *Artificial Intelligence in Engineering*, 3(3):125–140, 1988.
34. E. Davis. *Representations of Commonsense Knowledge.* Morgan Kaufmann Publishers, Inc., San Mateo, CA, 1990.
35. T. Dean and M. Wellman. *Planning and Control.* Morgan Kaufmann, 1991.
36. J. deKleer. Multiple representations of knowledge in a mechanics problem solver. In *Proceedings of IJCAI-77*, 1977.
37. J. deKleer and J. S. Brown. A qualitative physics based on confluences. In D. G. Bobrow, editor, *Qualitative Reasoning about Physical Systems*, pages 7–83. The MIT Press, Cambridge, MA, 1985.
38. M. diManzo and E. Trucco. Commonsense reasoning about flexible objects: A case study. In C. Mellish and J. Hallam, editors, *Advances in Artificial Intelligence.* 1987.
39. B. R. Donald. *Error Detection and Recovery for Robot Motion Planning with Uncertainty.* Ph.D. dissertation, Artificial Intelligence Laboratory, MIT, 1987.
40. D. Dubois and H. Prade. Order-of-magnitude reasoning with fuzzy relations. In *Proceedings of the IFAC Workshop on Advanced Information Processing in Automatic Control*, pages 195–200, Nancy, France, 1989.
41. M. R. Duffey and J. R. Dixon. Automating extrusion design: a case study in geometric and topological reasoning for mechanical design. In *Proceedings of the American Society of Mechanical Engineers (ASME) Computers in Engineering Conference*, San Francisco, CA, July 1988.
42. M. G. Dyer and M. Flowers. Toward automating design invention. In *Proceedings of AUTOFACT 6 Conference*, Anaheim, CA, 1984.
43. M. G. Dyer, M. Flowers, and J. Hodges. Edison: an engineering design invention system operating naively. *International Journal of Artificial Intelligence in Engineering*, 1(1):36–44, 1986.
44. M. Egenhofer and R. Franzosa. Point-set topological spatial relations. *International Journal of Geographical Information Systems*, 5(2):161–174, 1991.
45. H. A. ElMaraghy and W. R. Newcombe. Interactive kinematic analysis and synthesis of linkage mechanisms. In *Proceedings of the IEEE International Symposium on Mini and Micro Computers*, pages 179–182, Montréal, Québec, Nov. 1977.
46. H. A. ElMaraghy and S. Payandeh. Contact prediction and reasoning for compliant robot motions. *IEEE Transactions on Robotics and Automation*, 5(4):533–538, 1989.
47. A. G. Erdman. Computer aided design of mechanisms: 1984 and beyond. *Mechanism and Machine Theory*, 20(4):245–249, 1986.
48. M. Erdmann. Using backprojections for the fine motion planning with uncertainty. *The International Journal of Robotics Research*, 5(1):19–45, Spring, 1986.
49. B. Faltings. *Qualitative Place Vocabularies for Mechanisms in Configuration Space.* Ph.D. dissertation, Department of Computer Science, University of Illinois at Urbana-Champaign, 1987.
50. B. Faltings. Qualitative kinematics in mechanisms. *Artificial Intelligence*, 44(1–2):89–119, 1990.
51. B. Faltings. A symbolic approach to qualitative kinematics. *Artificial Intelligence*, 56(2–3):139–170, 1992.

52. B. Faltings. Qualitative kinematics and computer-aided design. In *Proceedings of the Second IFLP WG 5.2 Workshop on Intelligent CAD*, Cambridge, UK Sept., 1988.
53. B. Faltings, E. Baechler, and J. Primus. Reasoning about kinematic topology. In *Proceedings of IJCAI-89*, pages 1331–1336, 1989.
54. B. Faltings and P. Pu. Applying means-ends analysis to spatial planning. In *Proceedings of the 1991 IEEE/RSJ International Workshop on Intelligent Robots and Systems (IROS '91)*, pages 80–85, Osaka, Japan, Nov. 3–5 1991.
55. B. Faltings and K. Sun. Computer aided creative mechanism design. In *Proceedings of IJCAI-93*, pages 1451–1457, Aug. 1993.
56. B. Faverjon. Obstacle avoidance using an octree in the configuration space of a manipulator. In *Proceedings of the 1984 IEEE International Conference on Robotics and Automation*, Atlanta, March 1984.
57. B. Faverjon and P. Tournassoud. A local based approach for path planning of manipulators with a high number of degrees of freedom. In *Proceedings of the 1987 IEEE International Conference on Robotics and Automation*, pages 1152–1159, 1987.
58. B. Faverjon and P. Tournassoud. The mixed approach for motion planning: learning global strategies from a local planner. In *Proceedings of IJCAI-87*, volume 2, pages 1131–1137, Aug. 1987.
59. C. Ferrari and G. Chemello. Coupling fuzzy logic techniques with evidential reasoning for sensor data interpretation. In *Intelligent Autonomous Systems IAS-2*, pages 965–971, Amsterdam, December 11-14 1989.
60. S. Finger and J. R. Dixon. A review of research in mechanical engineering design. part i: descriptive, prescriptive, and computer based models of design processes. *Research in Engineering Design*, 1:51–67, 1989.
61. K. D. Forbus. Qualitative process theory. *Artificial Intelligence*, 24(1–3), 1984.
62. K. D. Forbus. Introducing actions into qualitative simulation. In *Proceedings of IJCAI-89*, pages 1273–1278, 1989.
63. K. D. Forbus, P. Nielsen, and B. Faltings. Qualitative kinematics: A framework. In *Proceedings of the International Joint Conference on Artificial Intelligence*, pages 430–435, 1987.
64. K. D. Forbus, P. Nielsen, and B. Faltings. Qualitative spatial reasoning: the `clock` project. *Artificial Intelligence*, 51(1–3):417–471, 1991.
65. D. H. Foster. Analysis of discrete internal representations of visual pattern stimuli. In Jacob Beck, editor, *Organization and Representation in Perception*, pages 319–341. Lawrence Erlbaum Associates, Hillsdale, NJ, 1982.
66. A. U. Frank. Qualitative spatial reasoning with cardinal directions. In *Proceedings of the Seventh Austrian Conference on Artificial Intelligence*, pages 157–167, Wien 1991.
67. P. Freedman and J. Liu. Using uncertain measurements to make plausible perceptual hypotheses: Application to autonomous navigation using sonar. In *Proceedings of the Third International Conference on Intelligent Autonomous Systems (IAS-3)*, Pittsburgh, PA, Feb. 1993.
68. C. Freksa. Qualitative spatial reasoning. In D. M. Mark and A. U. Frank, editors, *Cognitive and Linguistic Aspects of Geographic Space*. Kluwer, Dordrecht, 1991.
69. B. V. Funt. Problem solving with diagrammatic representations. *Artificial Intelligence*, 13:201–230, 1980.

70. J. S. Gero. Design prototypes: a knowledge representation schema for design. *AI Magazine*, 11(4):26–36, 1990.
71. J. S. Gero and M. A. Rosenman. A conceptual framework for knowledge-based design research at sydney university's design computing unit. *International Journal of Artificial Intelligence in Engineering*, 5(2):65–77, 1990.
72. B. Glavina. Solving findpath by combination of goal-directed and randomized search. In *Proceedings of the 1990 IEEE International Conference on Robotics and Automation*, pages 1718–1723, 1990.
73. S. Goss and J. L. Deneubourg. Harvesting by a group of robots. *Proceedings of the First European Conference on Artificial Life*, pages 195–204, Sydney, Australia, 1992. MIT Press/Bradford Books.
74. B. Grunbaum. *Convex Polytopes*. Wiley-Interscience, New York, 1967.
75. K. K. Gupta. Fast collision avoidance for manipulator arms: a sequential search strategy. In *Proceedings of the 1990 IEEE International Conference on Robotics and Automation*, pages 1724–1729, 1990.
76. K. K. Gupta and Z. Guo. Motion planning for many degrees of freedom: sequential search with backtracking. In *Proceedings of the IEEE International Conference on Robotics and Automation*, pages 2328–2333, Nice, France, May 1992.
77. H. W. Gusgen. Spatial reasoning based on allen's temporal logic. Technical Report TR-89-049, International Computer Science Institute, Berkeley, 1989.
78. B. Hajek. A tutorial survey of theory and applications of simulated annealing. In *Proceedings of the 24th IEEE Conference on Decision and Control*, pages 755–760, Lauderdale, FL, Dec. 1985.
79. B. Hajek. Cooling schedules for optimal annealing. *Mathematics of Operational Research*, 13(2):311–329, 1988.
80. R. S. Hartenberg and J. Denavit. *Kinematic Synthesis of Linkages*. McGraw-Hill, New York, NY, 1964.
81. A. Hayashi and B. Kuipers. Path planning for highly redundant manipulators using a continuous model. In *Proceedings of AAAI-91*, pages 666–672, 1991.
82. P. J. Hayes. The naive physics manifesto. In D. Michie, editor, *Expert Systems in the Microelectronic Age*. Edinburgh University Press, 1979.
83. P. J. Hayes. Naive physics 1: Ontology for liquids. In J.R. Hobbs and R.C. Moore, editors, *Formal Theories of the Commonsense World*. Ablex Publishing Corp., Norwood, NJ, 1985.
84. P. J. Hayes. The second naive physics manifesto. In J.R. Hobbs and R.C. Moore, editors, *Formal Theories of the Commonsense World*. Ablex Publishing Corp., Norwood, NJ, 1985.
85. V. Hayward. Fast collision detection schema by recursive decomposition of a manipulator workspace. In *Proceedings of the 1986 IEEE International Conference on Robotics and Automation*, pages 1044–1049, San Francisco, 1986.
86. P. Hebert, S. Betge-Brezetz, and R. Chatila. Decoupling odometry and exteroceptive perception in building a global world map of a mobile robot: The use of local maps. In *Proceedings of the IEEE International Conference on Robotics and Automation*, pages 757–764, Minneapolis, MN, 1996.
87. R. T. Hinkle. *Kinematics of Machines*. Prentice-Hall, New York, NY, 1953.
88. J. R. Hobbs and R. C. Moore. *Formal Theories of the Commonsense World*. Ablex Publishing, Norwood, NJ, 1985.
89. N. Hogan. Impedance control: an approach to manipulation: Parts 1-3. *Journal of Dynamic Systems, Measurement, and Control*, 107:1–24, March 1985.

90. K. H. Hunt. *Mechanisms and Motion.* John Wiley and Sons, New York, NY, 1959.
91. K. H. Hunt. *Kinematic Geometry of Mechanisms.* Oxford Engineering Science Series. Oxford University Press, Oxford, 1978.
92. Y. K. Hwang and N. Ahuja. A potential-field approach to path planning. *IEEE Transactions on Robotics and Automation*, 8(1):23–32, 1992.
93. J. Ilari and C. Torras. 2d path planning: a configuration space heuristic approach. *The International Journal of Robotics Research*, 9(1):75–91, 1990.
94. J. Ish-Shalom. The cs language concept: A new approach to robot motion design. *The International Journal of Robotics Research*, 4(1):42–58, 1985.
95. J. A. Janet, R. Gutierrez, T. A. Chase, M. W. White, and J. C. Sutton III. Autonomous mobile robot global self-localization using Kohonen and region-feature neural networks. *Journal of Robotic Systems*, 14(4):263–282, 1997.
96. L. Joskowicz. *Reasoning about Shape and Kinematic Function in Mechanical Devices.* Ph.D. dissertation, The Courant Institute of Mathematical Science, New York University, 1988.
97. L. Joskowicz and S. Addanki. From kinematics to shape: An approach to innovative design. In *Proceedings of the Second IFLP WG 5.2 Workshop on Intelligent CAD*, Cambridge, UK, Sept. 1988.
98. L. Joskowicz and E. Sacks. Incremental configuration space construction for mechanism analysis. In *Proceedings of AAAI-91*, pages 888–893, 1991.
99. G. Kanizsa. *Organization in Vision.* Praeger Publishers, New York, NY, 1979.
100. G. Kanizsa and W. Gerbino. Amodal completion: seeing or thinking. In Jacob Beck, editor, *Organization and Representation in Perception*, pages 167–190. Lawrence Erlbaum Associates, Hillsdale, NJ, 1982.
101. K. K. Gupta and S. W. Zucker. Planning smooth collision-free trajectories: Path, velocity and splines in free-space. Technical report, Centre for Intelligent Machines, McGill University, Montréal, 1986.
102. K. K. Gupta and S. W. Zucker. Toward efficient trajectory planning: the path-velocity decomposition. *The International Journal of Robotics Research*, 5(3):72–89, Fall 1986.
103. O. Khatib. Real-time obstacle avoidance for manipulators and mobile robots. In *Proceedings of the 1985 IEEE International Conference on Robotics and Automation*, pages 500–505, St. Louis, MO, May 1985.
104. O. Khatib. A unified approach to motion and force control of robot manipulators: The operation space formulation. *IEEE Journal of Robotics and Automation*, 3(1):43–53, 1987.
105. O. Khatib. Real-time obstacle avoidance for robot manipulators and mobile robots. *The International Journal of Robotics Research*, 5(1):90–98, Spring 1986.
106. H.-K. Kim. Qualitative kinematics of linkages. Technical Report UIUCDCS-R-90-1603, Computer Science Department, University of Illinois, Urbana, IL, 1990.
107. J.-O. Kim and P. K. Khosla. Real-time obstacle avoidance using harmonic potential functions. *IEEE Transactions on Robotics and Automation*, 8(3):338–349, 1992.
108. B. Kimia. *Toward a Computational Theory of Shape.* Ph.D. dissertation, Department of Electrical Engineering, McGill University, Montréal, 1991.

109. T. Kiriyama, Y. Ishida, T. Tomiyama, and H. Yoshikawa. Representation of behavior of design objects using qualitative physics. In *Proceedings of the Second IFIP W.G. 5.2 Workshop on Intelligent CAD*, San Francisco, CA, July 1988.
110. S. Kirkpatrick, C. D. Gelatt, and M. P. Vecchi. Optimization by simulated annealing. *Science*, 220:671–680, May 1983.
111. D. E. Koditshek. Robot planning and control via potential functions. In O. Khatib, J. J. Craig, and Tomás Lozano-Pérez, editors, *The Robotics Review 1*. The MIT Press, Cambridge, MA, 1989.
112. T. Kohonen. *Self-organization and Associative Memory,* 2nd Ed. New York: Springer-Verlag, 1988.
113. H. J. Kook and G. S. Novak. Representation of models for solving real-world physics problems. In *Proceedings of the IEEE Conference on Artificial Intelligence Applications*, 1990.
114. B. Kosko. *Neural Networks and Fuzzy Systems.* Prentice-Hall, 1992.
115. S. Kota. A qualitative matrix representation scheme for the conceptual design of mechanisms. In M. McCarthy, S. Derby, and A. Pisano, editors, *Mechanism Synthesis and Analysis*, pages 217–230. ASME, New York, NY, 1990.
116. S. Kota, A. G. Erdman, D. R. Riley, A. Esterline, and J. R. Slagle. A network based expert system for intelligent design of mechanisms. *AI EDAM*, 2(1):17–32, 1988.
117. G. A. Kramer. Solving geometric constraint systems. In *Proceedings of AAAI-90*, volume 2, pages 708–714, Boston, MA, Aug. 1990.
118. B. J. A. Krose and M. Eecen. A self-organizing representation of sensor space for mobile robot navigation. In *Proceedings of the IEEE/RSJ International Conference on Intelligent Robots and Systems*, pages 9–14, Munchen, Germany, 1994.
119. B. Kuipers. Modeling spatial knowledge. *Cognitive Science*, 2, 1978.
120. B. Kuipers. Commonsense reasoning about causality: Deriving behavior from structure. *Artificial Intelligence*, 24(1–3), 1984.
121. B. Kuipers. Qualitative simulation. *Artificial Intelligence*, 29:289–338, 1986.
122. B. Kuipers and Y.-T. Byun. A robot exploration and mapping strategy based on a semantic hierarchy of spatial representations. *Robotics and Autonomous Systems*, 8:47–63, 1991.
123. K. Kurumatani, T. Tomiyama, and H. Yoshikawa. Qualitative representation of machine behaviors for intelligent cad systems. *Mechanism and Machine Theory*, 25(3):325–334, 1990.
124. A. Kusiak. *Intelligent Manufacturing Systems.* Prentice Hall, Englewood Cliffs, NJ, 1990.
125. L. Latecki and R. Rohrig. Orientation and qualitative angle for spatial reasoning. In *Proceedings of IJCAI-93*, pages 1544–1549, Aug. 1993.
126. J.-C. Latombe. Towards automatic robot programming. In K. Rathmill, editor, *Control and Programming in Advanced Manufacturing*, pages 85–102. Springer-Verlag, London, UK, 1988.
127. J.-C. Latombe. *Robot Motion Planning.* Kluwer Academic Publishers, Norwell, MA, 1991.
128. J.-C. Latombe, A. Lazanas, and S. Shekhar. Robot motion planning with uncertainty in control and sensing. *Artificial Intelligence*, 52(1):1–47, 1991.
129. A. Lazanas and J.-C. Latombe. Landmark-based robot navigation. In *Proceedings of AAAI-92*, pages 816–822, 1992.

130. S. Y. Lee and F. J. Hsu. Picture algebra for spatial reasoning of iconic images represented in 2d c-string. *Pattern Recognition Letters*, 12:425–435, 1991.
131. D. Leven and M. Sharir. An efficient and simple motion-planning algorithm for a ladder moving in two-dimensional space amidst polygonal barriers. In *Proceedings of the First ACM Symposium on Computational Geometry*, pages 211–227, 1985.
132. J. Liu. *A Framework for the Qualitative Kinematics of Planar Mechanisms.* Technical Report, TR-CIM-90-14, Centre for Intelligent Machines, McGill University, Montréal, 1990.
133. J. Liu. *Autonomous Agents and Multi-Agent Systems.* World Scientific Publishing, Singapore, 2001.
134. J. Liu and L. K. Daneshmend. Qualitative physics for robot task planning: Part I, grammatical reasoning and commonsense augmentations. In *Proceedings of the 1991 IEEE/RSJ International Workshop on Intelligent Robots and Systems (IROS '91)*, Osaka, Japan, Nov. 3–5 1991.
135. J. Liu and L. K. Daneshmend. Qualitative physics for robot task planning: Part II, kinematics of mechanical devices. In *Proceedings of the 1991 IEEE/RSJ International Workshop on Intelligent Robots and Systems (IROS '91)*, Osaka, Japan, Nov. 3–5 1991.
136. J. Liu and L. K. Daneshmend. Qualitative analysis of task kinematics for compliant motion planning. In *Proceedings of the 1991 IEEE International Conference on Robotics and Automation*, pages 1258–1265, Sacramento, CA, April, 1991.
137. J. Liu and J. Wu. *Multi-Agent Robotic Systems.* CRC Press, Florida, USA, 2001.
138. T. Lozano-Pérez. Spatial planning: a configuration space approach. *IEEE Transactions on Computers*, C-32(2):108–109, 1983.
139. T. Lozano-Pérez. *An Approach to Automatic Robot Programming, AI Memo no. 842.* Artificial Intelligence Laboratory, MIT, 1985.
140. T. Lozano-Pérez. A simple motion-planning algorithm for general robot manipulators. *IEEE Journal of Robotics and Automation*, 3(3):224–238, 1987.
141. T. Lozano-Pérez, J. L. Jones, E. Mazer, P. A. O'Donnell, and Eric L. Grimson. Handey: A task-level robot system. In R. C. Bolles, editor, *Robotics Research: The Fourth International Symposium*, pages 29–36. The MIT Press, Cambridge, MA, 1988.
142. T. Lozano-Pérez, M. T. Mason, and R. H. Taylor. Automatic synthesis of fine-motion strategies for robots. *The International Journal of Robotics Research*, 3(1):3–24, Spring, 1984.
143. T. Lozano-Pérez and M. A. Wesley. An algorithm for planning collision-free paths among polyhedral obstacles. *Communications of the ACM*, 22(10):560–570, 1979.
144. V. J. Lumelsky. Effect of kinematics on motion planning for planar robot arms moving amidst unknown obstacles. *IEEE Journal of Robotics and Automation*, 3(3):207–223, 1987.
145. V. J. Lumelsky and A. A. Stepanov. Dynamic path planning for a mobile automaton with limited information on the environment. *IEEE Transactions on Automatic Control*, AC-31, 1986.
146. M. Lundy and A. Mees. Convergence of an annealing algorithm. *Mathematical Programming*, 34:111–124, 1987.

147. M. T. Mason. Compliance and force control for computer controlled manipulators. *IEEE Transactions on Systems, Man, and Cybernetics*, 11(6):418–432, 1981.
148. M. Mataric. A distributed model for mobile robot environment-learning and navigation. Technical Report 1228, AI Laboratory, MIT, April 11 1990.
149. M. L. Mavrovouniotis and G. Stephanopoulos. Reasoning with order of magnitude and approximate relations. In *Proceedings of AAAI-87*, pages 626–630. AAAI Press/The MIT Press, 1987.
150. E. Mazer, J. Pertin-Troccaz, and J.-M. Lefevre. Act: a robot programming environment. In *Proceedings of the 1991 IEEE International Conference on Robotics and Automation*, pages 1427–1432, Sacramento, CA, April 1991.
151. D. McDermott and E. Davis. Planning routes through uncertain territory. *Artificial Intelligence*, 22(2):107–156, 1984.
152. N. Metropolis, A. Rosenbluth, M. Rosenbluth, A. Teller, and E. Teller. Equations of state calculations by fast computing machines. *Journal of Chem. Phys.*, 21:1087–1091, 1953.
153. T. M. Mitchell, M. T. Mason, and A. D. Christiansen. Toward a learning robot. Technical Report CMU-CS-89-106, Computer Science Department, Carnegie Mellon University, Pittsburgh, PA, 1989.
154. P. Molitor. Layer assignment by simulated annealing. *Microprocessing and Microprogramming*, 16(4–5):345–349, 1985.
155. R. De Mori and R. Prager. Perturbation analysis with qualitative models. In *Proceedings of IJCAI-89*, pages 1180–1186, 1989.
156. A. Mukerjee. Accidental alignments: An approach to qualitative vision. In *Proceedings of the IEEE International Conference on Robotics and Automation*, pages 1096–1101. Apr., 1991.
157. N. H. Narayanan and B. Chandrasekaran. Reasoning visually about spatial interactions. In *Proceedings of IJCAI-91*, pages 360–365, 1991.
158. P. P. Nayak, L. Joskowicz, and S. Addanki. Automated model selection using context-dependent behaviors. In *Proceedings of AAAI-92*, pages 710–716, 1992.
159. C. OĎuńlaing, M. Sharir, and C.-K. Yap. Retraction: A new approach to motion planning. In *Proceedings of the 15th FOCS*, pages 207–220, 1983.
160. C. H. Orgill. An online mobility model for robotic assemblages using orthogonal physical constraints. In *Proceedings of the International Workshop on Artificial Intelligence for Industrial Applications*, pages 401–410, 1988.
161. D. Pagac, E. M. Nebot, and H. Durrant-Whyte. An evidential approach to probabilistic map-building. In *Proceedings of the 1996 IEEE International Conference on Robotics and Automation*, pages 745–750, Minneapolis, Minnesota, USA, 1996.
162. R. Prager, P. Belanger, and R. De Mori. A knowledge-based system for troubleshooting real-time models. In L. E. Widman, K. A. Loparo, and N. R. Nielsen, editors, *Artificial Intelligence, Simulation, and Modeling*, pages 511–543. John Wiley and Sons, New York, NY, 1989.
163. P. Pu. *Qualitative Simulation of Ordinary and Intermittent Mechanisms*. Ph.D. dissertation, Department of Computer and Information Science, University of Pennsylvania, 1989.
164. O. Raiman. Order of magnitude reasoning. In *Proceedings of AAAI-86*, pages 100–104, 1986.

165. D. Randell. Analyzing the familiar: a logical representation of space and time. In *Proceedings of the Third International Workshop on Semantics of Time, Space, and Movement*, Toulouse, 1991.
166. F. Reuleaux. *The Kinematics of Machinery: Outlines of a Theory of Machines, 1876.* Reprinted by Dover Publications Inc., New York, NY, 1963.
167. I. Rock. *The Logic of Perception.* The MIT Press, Cambridge, MA, 1983.
168. K. D. Rueb and K. C. Wong. Structuring free-space as a hypergraph for roving robot path planning and navigation. *IEEE Transactions on Pattern Analysis and Machine Intelligence*, PAMI-9(2):263–273, 1987.
169. N. H. Salem and S. Manoochehri. Autocad based computer-aided mechanism design system. In *Cams, Gears, Robot and Mechanism Design: Proceedings of the 1990 ASME Design Technical Conferences – 21st Biennial Mechanisms Conference*, Chicago, IL, Sept. 1990.
170. G. N. Sandor and A. G. Erdman. *Advanced Mechanism Design: Analysis and Synthesis.* Prentice-Hall, Englewood Cliffs, NJ, 1984.
171. J. T. Schwartz and M. Sharir. On the piano movers' problem I: The case of a two-dimensional rigid polygonal body moving amidst polygonal barriers. *Communications on Pure and Applied Mathematics*, 36:345–398, 1983.
172. J. E. Shigley and J. J. Uicker. *Theory of Machines and Mechanisms.* McGraw-Hill,New York, NY, 1980.
173. Y. Shoham. Naive kinematics: One aspect of shape. In *Proceedings of the International Joint Conference on Artificial Intelligence*, pages 436–442. 1985.
174. P. Siarry, L. Bergonzi, and G. Dreyfus. Thermodynamic optimization of block placement. *IEEE Transactions on Computer-Aided Design*, CAD-6(2):211–221, 1987.
175. R. Simmons. "commonsense" arithmetic reasoning. In *Proceedings of AAAI-86*, pages 118–124, 1986.
176. C. Stanfill. The decomposition of a large domain: Reasoning about machines. In *Proceedings of the First National Conference on Artificial Intelligence*, pages 387–390. 1983.
177. D. Subramanian. Conceptual design and artificial intelligence. In *Proceedings of IJCAI-93*, pages 800–809, Aug. 1993.
178. C. H. Suh and C. W. Radcliffe. *Kinematics and Mechanisms Design.* Wiley and Sons, New York, NY, 1978.
179. H. Takeda, P. Veerkamp, T. Tomiyama, and H. Yoshikawa. Modeling design process. *AI Magazine*, 11(4):37–48, 1990.
180. T. Tomiyama, T. Kiriyama, H. Takeda, D. Xue, and H. Yoshikawa. Metamodel: A key to intelligent cad. *Research in Engineering Design*, 1:19–34, 1989.
181. L. Travé-Massuyès, N. Piera, and A. Missier. What can we do with qualitative calculus today. In *Proceedings of the IFAC Workshop on Advanced Information Processing in Automatic Control*, pages 207–212, Nancy, France, 1989.
182. S. Udupa. *Collision Detection and Avoidance in Computer Controlled Manipulators.* Ph.D. dissertation, Department of Electrical Engineering, California Institute of Technology, 1977.
183. J. J. Uicker, J. Denavit, and R. S. Hartenberg. An iterative method for the displacement analysis of spatial mechanisms. *Journal of Applied Mathematics, Transactions of the ASME*, June, 1964.
184. K. T. Ulrich. *Computation and Pre-Parametric Design.* Ph.D. dissertation, MIT Artificial Intelligence Laboratory, 1988.

185. K. T. Ulrich and W. P. Seering. Synthesis of schematic descriptions in mechanical design. *Research in Engineering Design*, 1:3–18, 1989.
186. J. Vandorpe, H. Van Brussel, and H. Xu. Exact dynamic map-building for a mobile robot using geometrical primitives produced by a 2D ranger finder. In *Proceedings of the IEEE International Conference on Robotics and Automation*, pages 901–908, Minneapolis, MN, 1996.
187. D. S. Weld. Combining discrete and continuous process models. In *Proceedings of IJCAI-85*, Los Angeles, CA, Aug. 1985.
188. S. H. Whitesides. Computational geometry and motion planning. In G. T. Toussaint, editor, *Computational Geometry*, pages 377–427. Elsevier Science Publishers B. V., North-Holland, 1985.
189. D. Whitney. Historical perspective and state of the art in robot force control. *The International Journal of Robotics Research*, 6(1):3–14, 1987.
190. B. C. Williams. Qualitative analysis of mos circuits. *Artificial Intelligence*, 24(1–3), 1984.
191. P. H. Winston. *Artificial Intelligence*. Addison-Wesley Publishing, Reading, MA, 1984.
192. B. Yang, U. Datta, P. Datseris, and Y. Wu. An integrated system for design of mechanisms by an expert system – domes. *AI EDAM*, 3(1):53–70, 1989.
193. C.-K. Yap. *Coordinating the motion of several discs*. Robotics Lab., NYU-Courant Institute, no. 16, 1984.
194. C.-K. Yap. Algorithmic motion planning. In J. T. Schwartz and C.-K. Yap, editors, *Algorithmic and Geometric Aspects of Robotics*, volume 1, pages 95–143. Lawrence Erlbaum Associates, Publishers, Hillsdale, NJ, 1987.
195. C.-K. Yap. How to move a chair through a door. *IEEE Journal of Robotics and Automation*, 3(3):172–181, 1987.
196. L. A. Zadeh. Fuzzy logic. *IEEE Computer*, April:83–93, 1988.
197. D. Zhu and J.-C. Latombe. Mechanization of spatial reasoning for automatic pipe layout design. *AI EDAM*, 5(1):1–20, 1991.
198. D. Zhu and J.-C. Latombe. New heuristic algorithms for efficient hierarchical path planning. *IEEE Transactions on Robotics and Automation*, 7(1):9–20, 1991.

Index